Oliver Näckel

Development of an Air Coil Superconducting Fault Current Limiter

Karlsruher Schriftenreihe zur Supraleitung Band 017

HERAUSGEBER

Prof. Dr.-Ing. M. Noe
Prof. Dr. rer. nat. M. Siegel

Eine Übersicht über alle bisher in dieser Schriftenreihe
erschienene Bände finden Sie am Ende des Buches.

Development of an Air Coil Superconducting Fault Current Limiter

by
Oliver Näckel

KIT Scientific Publishing

Dissertation, Karlsruher Institut für Technologie (KIT)
Fakultät für Elektrotechnik und Informationstechnik, 2015
Hauptreferent: Prof. Dr.-Ing. Mathias Noe
Korreferent: Prof. Dr. Frédéric Sirois

Impressum

KIT Scientific Publishing

Karlsruher Institut für Technologie (KIT)
KIT Scientific Publishing
Straße am Forum 2
D-76131 Karlsruhe

KIT Scientific Publishing is a registered trademark of Karlsruhe Institute of Technology. Reprint using the book cover is not allowed.

www.ksp.kit.edu

Print on Demand 2016

ISSN 1869-1765
ISBN 978-3-7315-0526-6
DOI 10.5445/KSP/1000054588

Development of an Air Coil Superconducting Fault Current Limiter

Zur Erlangung des akademischen Grades eines
DOKTOR-INGENIEURS
von der Fakultät für
Elektrotechnik und Informationstechnik
des Karlsruher Instituts für Technologie (KIT)
genehmigte
DISSERTATION
von
Dipl.-Ing. Oliver Näckel
geb. in: Balingen

Tag der mündlichen Prüfung: 1. Dezember 2015
Hauptreferent: Prof. Dr.-Ing. Mathias Noe
Korreferent: Prof. Dr. Frédéric Sirois

Acknowledgements

This work was created at the Institute for Technical Physics at the Karlsruhe Institute of Technology, which provided an excellent and professional environment.

I would like to thank Prof. Dr.-Ing. Mathias Noe for his supervision, continuous effort and support in order to ensure the scientific quality of this work as well as encouraging me to participate conferences and support for my research visit. I also would like to thank Prof. Frédéric Sirois for his interest in this work and to undertake the second review.

For their professional and technical assistance during the experiments and measurements at the SET laboratory I would like to thank Andrej Kudymov, Severin Strauss, Johann Wilms and Uwe Walschbuger as well as Jürgen Hieringer for his help performing preliminary experiments.

The power hardware-in-the-loop tests would not have been possible with the support of the Karlsruhe House of Young Scientitsts (KHYS) and the CAPS team at the Florida State University, especially James Langston, Michael Steurer and Lukas Graber.

Zusammenfassung

Das elektrische Energienetz ist ein existentieller Bestandteil heutiger Infrastruktur. Die kontinuierliche und unterbrechungsfreie Versorgung mit elektrischer Energie ist grundlegend für Produktion, Kommunikation, Transport und unser alltägliches Leben. Die Erzeugung elektrischer Energie und der Verbrauch müssen zu jedem Zeitpunkt übereinstimmen, damit sich das Energienetz in Balance befindet. Fehlerströme und insbesondere Kurzschlussströme können diese Balance nachhaltig stören, sowie unterschiedliche und weitreichende Folgen für das elektrische Energienetz haben und die Stabilität des Netzes gefährden. Die Folgen können von kurzen Unterbrechungen bis hin zur Zerstörung von elektrischen Betriebsmitteln wie Transformatoren oder Generatoren reichen.

Es ist daher unumgänglich Maßnahmen zu ergreifen, um Fehlerströme zu begrenzen. Neben etablierten Betriebsmitteln, wie Sicherungen und Drosselspulen, haben sich supraleitende Strombegrenzer zu einem kommerziell erhältlichen Betriebsmittel zur Kurzschlussstrombegrenzung entwickelt. Supraleitende Strombegrenzer begrenzen Kurzschlussströme zuverlässig und schnell (innerhalb der ersten Halbwelle) und bieten darüber hinaus die Möglichkeit nach Begrenzung des Fehlerstromes die Energieübertragung fortzusetzen („recovery under load").

Es existieren verschiedene Typen von supraleitenden Strombegrenzern, die Gegenstand von Forschung und Entwicklung sind. Diese Arbeit beschreibt die Funktionsweise und den Aufbau eines „Air Coil Superconducting Fault Current Limiter" (AC-SFCL), sowie die Auslegung, den Entwurf und Test eines 60 kVA, 400 V, z = 6% AC-SFCL Demonstrators. Der AC-SFCL ist ein supraleitender Strombegrenzer mit induktiver Widerstandseinkopplung. Hierzu werden zwei Solenoidwicklungen mit gleicher Bauhöhe konzentrisch angeordnet bzw. ineinander gestellt. In einer Wicklung aus nicht-supraleitendem Material, der Primärwicklung, fließt der Transportstrom bei Raumtemperatur. Diese wird von der zweiten, supraleitenden Wicklung, der Sekundärwicklung, abgeschirmt. Im Normalbetrieb besitzt der AC-SFCL durch die Abschirmung eine sehr niedrige Impedanz. Im Begrenzungsfall geht die supraleitende Sekundärwicklung in den normalleitenden Bereich über und entwickelt einen großen Widerstand. Damit steigt die Impedanz des AC-SFCL und begrenzt effektiv den Kurzschlussstrom.

In dieser Arbeit werden zwei Anwendungsfälle unterschieden. Eine Möglichkeit besteht darin eine Drosselspule mit einem supraleitenden Einsatz nachzurüsten. Dieses als Retrofit bezeichnete Vorgehen wertet die Drosselspule zu einem AC-SFCL auf. Im Normalbetrieb ist die Impedanz der Drosselspule so weit reduziert, dass sie für das Netz nahezu unsichtbar wird. Dies minimiert die Netzrückwirkungen der Drosselspule und verbessert die Stabilität des Netzes. Tritt ein Fehlerstrom auf wird der Supraleiter im

Einsatz normalleitend und die Impedanz steigt. Damit wird der Fehlerstrom effektiv begrenzt.

Die zweite Möglichkeit besteht darin den AC-SFCL von Grund auf neu zu entwerfen. Hierbei wird die Geometrie der Primärwicklung derart optimiert werden, dass die Impedanz im Normalbetrieb weiter sinkt im Vergleich zum Retrofit. Die Impedanz und Strombegrenzung im Fehlerfall bleiben dabei erhalten.

Um einen AC-SFCL auszulegen werden die nötigen Entwurfsgleichungen aufgestellt und in einem Entwurfsgang zusammengefasst. Mit Hilfe dieses Entwurfsganges wird dann ein 60 KVA, 400 V, z = 6% AC-SFCL Demonstrator berechnet, konstruiert und anschließend gebaut. Durch Kurzschlussversuche mit dem Demonstrator konnte die Funktionsweise untersucht, das Prinzip des Begrenzers belegt und die Ergebnisse des Entwurfsganges verifiziert werden.

Weiterführende experimentelle Untersuchungen mit „Power Hardware-in-the-loop" (PHIL) bestätigen die Messergebnisse der Kurzschlussversuche und bescheinigen dem AC-SFCL die gleiche Strombegrenzungsfähigkeit bei symmetrischen und unsymmetrischen Kurschlüssen in einem dreiphasigen System.

Abschließend werden mit den Entwurfsgleichungen konzeptionelle AC-SFCLs Designs von Begrenzern für Mittel-, Hoch-, und Höchstspannung vorgestellt und insbesondere der Retrofit mit einem Entwurf mit neu konzipierter Primärwicklung verglichen.

Table of Contents

1 Introduction, motivation and scope of work

Generation, transfer and distribution of electrical energy is a vital task for the infrastructure, industry and our daily lives. The demand of electrical energy must be satisfied continuously without any interruption and the balance between generation and consumption must be well preserved. Fault currents can severely harm this stable condition of the power grid, because they can cause power outages and even blackouts. They pose a vital threat for the power equipment, like transformers and generators and replacing such power equipment can be costly. An interruption of the supply with electric energy caused by a fault current can lead to costs, which are significantly higher than the value of the actual delivered power. These costs mainly depend on interruption duration, interruption time (e.g. nighttime or daytime) and affected customers (e.g. residential or industrial). The costs of power outages ("value of lost load") have been assessed by several studies in Europe and the US for different scenarios [HE06], [NKB07], [BH09], [LT11], [PHE11], [RSS12], [ZP12], [RIA13].

Fault currents have various reasons, with the short-circuit currents, which can easily reach 20 times the nominal current, probably being the most severe ones. Especially in growing urban areas with high power densities installed devices and measures to limit short-circuit currents reach their technical limits. For this reason superconducting fault current limiters are a viable option to cope with the challenge to effectively limit short-circuit currents at increasing short-circuit capacity.

Established and implemented measures differ in behavior and applicability on different voltage levels. For low and medium voltage levels, fuses are a conventional and common measure to limit fault currents. They react fast (within the first half cycle), self-triggered and limit fault currents reliably. The disadvantages are the necessity to change the fuse after the fault and the limitation to medium voltage. Air core reactors are a common and commercially available measure to limit fault currents on all voltage levels. In principle an air core reactor is a coil made of copper or aluminium, limiting the current due to their impedance, which is mainly reactive. This impedance however does not only take effect during fault current limitation, but also during normal operation. The result is a voltage drop and reactive losses, which needs to be compensated, normally by feeding with a higher voltage. As a consequence air core reactors cannot be applied with any desired impedance, otherwise the reactive voltage drop and the decreased power transfer capability may considerably decrease the system stability and cause system perturbations. Therefore the voltage drop of an air core reactor usually ranges from 3% to 10% of the rated voltage.

Superconducting fault current limiters (SFCLs) have been field tested and are on the verge of becoming a commercial and accepted measure to limit fault currents. In general

SFCLs offer fast (within the first half cycle) and reliable current limitation, while maintaining a low impedance during normal operation. SFCLs can not only limit a fault current, but also allow the continuous transmission of power after the fault (recovery under load). Possible applications in power grids have been identified, investigated [SID95], [BSS97], [NO99], [LBS03], [KYT05] and several different types of SFCLs are subject to research and development [Mor13], [EPRI09], [NS07], [NP97], [GR93]. Each of these types implements a different approach to limit fault currents (e.g. resistive or inductive fault current limitation), which implies possible advantages and disadvantages.

This work proposes the air coil superconducting fault current limiter (AC-SFCL) as a measure to effectively limit fault currents. The AC-SFCL aims to overcome the disadvantages of the air core reactor by retrofitting it with a secondary, short-circuited superconducting winding. The transport current is conducted through the air core reactor, which acts as a primary winding and induces a current in the secondary superconducting winding. During normal operation the secondary winding shields the primary winding and lowers the impedance significantly. If a fault current occurs the induced current quenches the secondary superconducting winding, generates a resistance and the shielding collapses. The result is a combined resistive-inductive limitation of the fault current. The mechanism in the secondary winding is essentially the same as in the resistive type SFCL (the transition from superconducting to normal state). However since the windings of the AC-SFCL are inductively coupled the mutual inductance of both coils limits the fault current as well. The inductive coupling introduces another advantage: No current leads are necessary for the operation. This reduces the losses tremendously, and only the AC losses of the *RE*BCO tapes and the losses in the primary winding as well as the heat transfer through the cryostat contribute to the losses during operation.

The main objectives of this work are:

- Develop the necessary theoretical background and equations for the design of the AC-SFCL. Compile the equations into a design method.
- Design an AC-SFCL demonstrator and perform short-circuit tests to verify and prove the concept of the limiter.
- Validate the calculated results with the measurements and improve the design process, if necessary.
- The obtained results serve as a foundation for conceptual designs of AC-SFCLs for different voltage levels.

The fundamentals of technical superconductors relevant for the application in a SFCL and the principle of the AC-SFCL are presented in chapter 2. The definitions of the critical values of a superconductor and the structure of *RE*BCO conductors, which are closely connected to the electrical parameters of the AC-SFCL are important for the operation and understanding of the AC-SFCL. The chapter also describes common measures for fault current limitation and the state-of-the-art of SFCLs.

Chapter 3 focuses on the derivation of the design equations. A key role in the calculation of the AC-SFCL is the development of design equations. These design equations allow the calculation of the electrical parameters depending on the geometrical parameters. In order to describe the electrical behavior, an equivalent circuit diagram is deployed, which is derived from the transformer design and incorporates the properties of superconducting *RE*BCO tapes. Since AC-losses are one important loss mechanism, equations are presented for their calculation. The design equations are ordered in such a manner, that the AC-SFCL can be calculated automatically by specifying the deployment (voltage level, rated power, etc.), variation of geometrical parameters and the properties of the *RE*BCO tapes. Hereby two possible applications are addressed independently: The retrofit of a commercially available air core reactor and the design with an improved primary winding in order to improve the electrical parameters, namely the ratio between impedance during fault and during normal operation.

The equations and the design method are used to design a S = 60 kVA, V = 400 V and z = 6% AC-SFCL demonstrator. Chapter 4 describes the design of the demonstrator and the measurements performed. The preliminary investigations included characterization of the *RE*BCO tapes, manufacturing a small test setup as well as soldering and quenching of single superconducting short-circuited *RE*BCO rings. Using the equations presented in chapter 3 the AC-SFCL demonstrator is designed meeting the specification and at the same time keeping a compact geometry. With the final design a simulation was performed to estimate the current limiting capability. First load measurements were performed to verify the expected significant lower impedance of the AC-SFCL demonstrator during load compared to the air core reactor (primary winding only). After the normal operation was confirmed, short-circuit test have been performed and the current limiting capability was investigated. Therefore, short-circuit measurements have been performed and were varied in terms of prospective current, fault duration and fault angle. The results are compared to the simulation and the calculated steady-state design values in order to prove the concept of the AC-SFCL and the design approach.

Additional tests were performed in a power hardware-in-the-loop system (PHIL) after the successful short-circuit tests of the AC-SFCL. Chapter 5 describes the basics of such a PHIL system, which is a combination of the measurement of a single power device, such as the AC-SFCL demonstrator, and a real-time simulation of a surrounding power grid for example. This allowed to investigate the performance of the AC-SFCL in real power grids for symmetrical and unsymmetrical short-circuits. A major challenge was to find an interface algorithm, which represents the AC-SFCL in the simulated environment correctly, delivers reliable results and enables a stable operation of the system. Once a suitable interface algorithm was found, a three phase system with a transient model of a generator was implemented in the real-time simulation environment. One of the phases was connected to the power hardware and the physical AC-SFCL demonstrator, while the other phases were connected to the simulated AC-SFCLs. The AC-SFCL demonstrator was subject to real fault currents and the generated impedance during a fault was fed back into the simulated system and applied to the simulated AC-SFCLs, if subject to the fault as well.

Future prospects of AC-SFCLs are explored in chapter 6 with conceptual designs for medium, high and ultra-high voltage. The objective of the conceptual designs is to examine the feasibility and advantages of the AC-SFCL for different voltage levels in general. For each voltage level the retrofit of an air core reactor is compared to a design with improved primary winding in terms of impedance during normal operation and fault limiting operation, losses and superconductor demand. Hereby the influence of the major parameters of the *RE*BCO tape and the geometrical parameters of the primary winding (for the improved design) are varied and their influence on the impedances is explored.

The results of this work are summarized in chapter 7 along with future prospects.

2 Fundamentals of superconductors for applications and state-of-the-art fault current limitation

Superconductors include a variety of elements and compounds. The first superconducting material, mercury, was discovered by Heike Kamerlingh Onnes in 1911 [Onn12]. He discovered, that mercury had no measureable resistance at a temperature of T = 4.15 K. The non-measureable resistance below a certain temperature, the so called critical temperature T_c, is one criterion to determine, if a material or compound is considered a superconductor. The second criterion is the behavior in magnetic fields, in which a superconductor shows ideal diamagnetic behavior while in superconducting state [MO12]. Exposed to an increasing external magnetic field two types of superconductors can be distinguished. In type I superconductors the external magnetic field penetrates the superconductor abruptly and the superconductor transits to normal state, if the external magnetic field exceeds the critical magnetic field H_{c1}. In type II superconductors the magnetic field penetrates the superconductor in flux vortices exposed to magnetic fields higher than the critical magnetic field H_{c1} [Ber87]. This penetration continues until the superconductivity breaks down at the critical magnetic field H_{c2}. The critical current I_c is the third critical value of a superconductor. For transport currents above the critical current I_c the superconducting state vanishes. The critical values are described in more detail in chapter 2.1.1.

Only a few of the discovered superconducting materials have been found suitable for industrial applications. The so called low temperature superconductors (LTSCs), namely Niobium-Titanium (NbTi) and Niobium-Tin (Nb_3Sn), are relevant for magnet applications such as magnet resonance imaging (MRI), nuclear magnetic resonance (NMR), accelerator magnets and magnets for fusion reactors. The critical temperature of NbTi is T_c = 9 K and of Nb_3Sn is T_c = 18 K. A common coolant for operation of the LTSCs is liquid helium (4.15 K), which requires a large cooling power and therefore high investment and operational costs. This is a substantial disadvantage and eliminates LTSCs as an option for applications in the field of electrical power engineering.

In 1986 the Ba-La-Cu-O was discovered by Bednorz and Müller with a critical temperature of T_c = 35 K [Bed86] . This discovery was followed by the discovery of $YBa_2Cu_3O_{7-\delta}$ (T_c = 92 K) in 1987 and $Bi_2Sr_2Ca_1CuO_y$ (T_c = 92 K) and $Bi_2Sr_2Ca_2Cu_3O_y$ (T_c =110 K) in 1988 [WAT87], [Mae88]. The crucial advantage of these materials is the possibility to cool them with liquid nitrogen. Liquid nitrogen boils at T = 77.4 K under ambient pressure, which allows cost-effective cooling and makes them an attractive option for applications in the field of electrical power engineering. BSCCO bulk material could be successfully applied for SFCLs [EBW01], [EBN03]. Wires or tapes based on

BSSCO are not suitable for resistive type SFCLs, since the BSCCO material is embedded in a highly conductive silver matrix (about 70% of the cross-section of the wire), which generates an insufficient resistance during fault operation. For the application in the AC-SFCL only high temperature superconductors (HTS) based on *RE*BCO conductors were considered. In this work the *RE* stands for rare earth, which includes materials as Yttrium, Dysprosium or Gadolinium. The major properties relevant for applications of *RE*BCO are described below.

2.1 High temperature superconductors for applications

2.1.1 Critical values of technical superconductors

The superconducting state of a superconductor can only be maintained, if temperature, current and magnetic field remain below their corresponding critical values. Figure 2.1 shows qualitatively the resistance depending on temperature for a conventional conductor and a superconductor as well as three definitions of the critical temperature T_c.

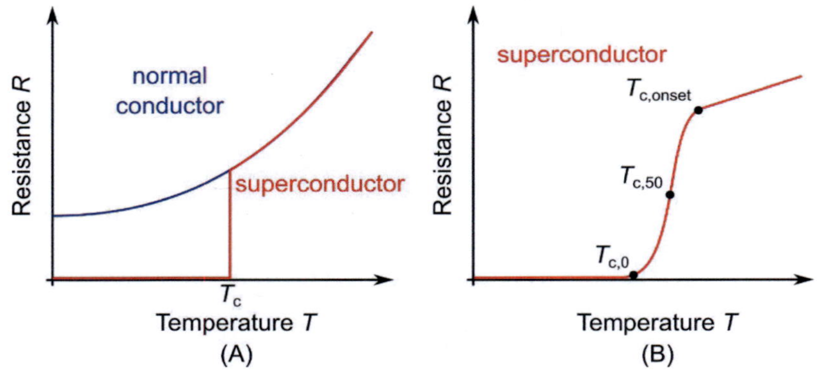

Figure 2.1: (A) Resistance depending on temperature for superconductors and normal conductors and (B) magnification around the transition and different definitions of the critical temperature [Cav98]

In case of the conventional conductor the resistance decreases with temperature until at a residual resistance is remaining at $T = 0$ K. The resistance R of a superconductor shows a decrease with decreasing temperature T until the so called critical temperature T_c is reached. At $T = T_c$ the resistance drops almost instantaneously to non-measurable values and the superconductor is considered to be in superconducting state. As long as the temperature remains below T_c the superconductor remains in superconducting state. If the temperature rises again and exceeds critical temperature T_c the superconductor becomes normal conducting again. This transition from superconducting to normal state is called a quench. This transition is not ideally vertical and several definitions for the critical temperature exist [Cav98]:

- the temperature $T_{c,onset}$, where the superconductor first starts to show highly non-linear behavior
- the temperature $T_{c,50}$ at which the electrical resistivity is dropped to 50% of the normal conducting regime
- the critical temperature $T_{c,0}$, where the electrical resistivity has measurable the first time coming from lower temperatures

In this work the definition of the critical temperature $T_{c,50}$ is used.

Superconductors in superconducting state can carry a substantial higher current density than normal conductors like copper [LGFP01]. The current carrying capability of a superconductor in superconducting state is limited by its critical current I_c or critical current density j_c respectively. Figure 2.2 (A) shows the characteristic dependency of current density on an electric field E.

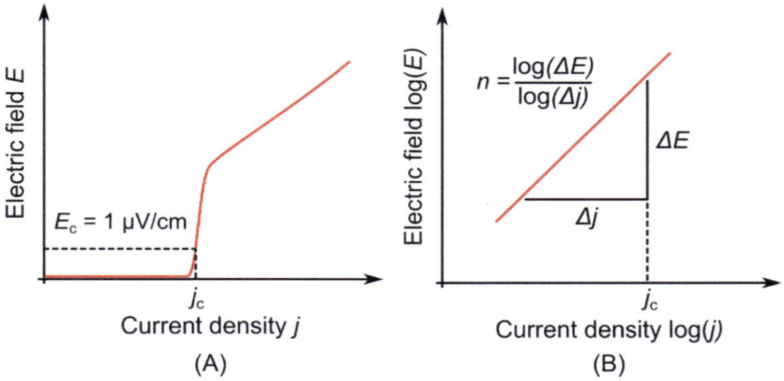

Figure 2.2: (A) Definition of the critical current density j_c and (B) definition of the n-value

The transition from superconducting to normal state is not occurring instantaneously, but can be described by the so called power law:

$$E = E_c \cdot \left(\frac{I}{I_c}\right)^n \tag{2.1}$$

Herein the n-value is a material specific parameter. The definition of the n-value is shown in Figure 2.2 (B). Typical n-values for *RE*BCO conductors are in the order of up to 40 [SCX08]. In commercial available *RE*BCO tapes the critical current I_c is not uniform along the tape length, but deviates within a few percent [ZLF11].

Superconductors displace the magnetic field outwards and show ideal diamagnetic behavior. Once the magnetic flux density exceeds the critical flux density B_c of the superconductor the magnetic field penetrates the superconductor and eventually the superconducting state collapses. Superconductors can be classified in two different types regarding the critical magnetic field. The behavior of a so called type I superconductor is shown in Figure 2.3 (A). If a type I superconductor is exposed to a magnetic field, it displaces the magnetic field until the critical magnetic flux density B_{c1} is reached. At B_{c1} the superconducting state collapses instantaneously and the magnetic field penetrates the superconductor completely. The superconductor is then in the normal phase. The critical magnetic flux density B_{c1} of type I is exceeded by the magnetic field of small transport currents. This makes type I superconductors ineligible for technical applications.

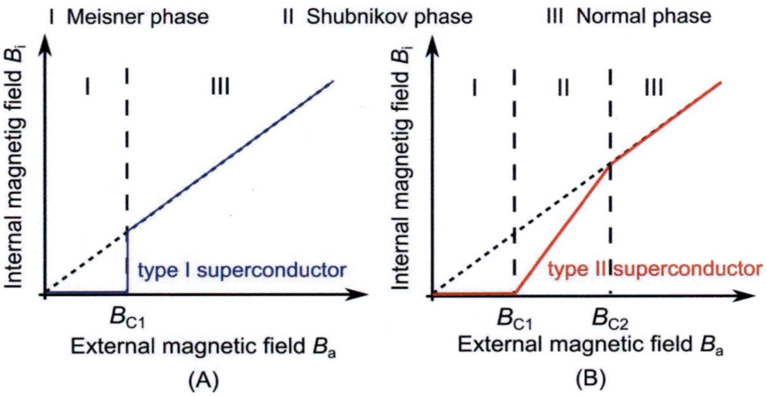

Figure 2.3: Behavior in an external magnetic field of (A) type I superconductors and (B) type II superconductors

The behavior under exposure of an external magnetic field of type II superconductors is shown in Figure 2.3 (B). If the magnetic flux density B_a exceeds the critical magnetic field B_{c1} the magnetic field starts to penetrate the superconductor in quantified flux lines. These flux lines are vortices with a normal conducting center. The magnetic field starts penetrating the superconductor from the outside to the center. With increasing magnetic field the number of flux lines is increasing as well until the superconductor is fully penetrated at the magnetic flux density B_{c2}. In magnetic fields above B_{c2} the superconductor is in the normal phase. The transitional region between B_{c1} and B_{c2} is called Shubnikov phase. Generally, all technical superconductors are type II superconductors, which are operated in the Shubnikov phase.

For the operation of superconductors in applications, such as superconducting fault current limiters, it is necessary to ensure, that the temperature, current and magnetic field do not exceed their respective critical values, otherwise the superconducting state will break down immediately. Figure 2.4 shows the combination of all three critical parameters according to [Kom95].

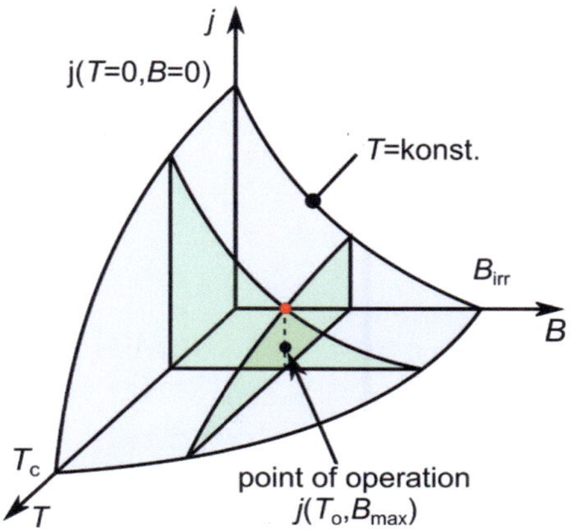

Figure 2.4: Combination of all three critical values to determine the point of operation with safety margins according to [Kom95]

The critical values are actually depending on each other. The critical magnetic flux density B_{c2} can be approximated by [Kom95]

$$B_{c2} = B_{c2}(T = 0) \cdot \left[1 - \left(\frac{T}{T_c}\right)^2\right] \tag{2.2}$$

Accordingly the critical current density j_c can be expressed depending on the magnetic flux density B at constant temperature T [KHS63]

$$j_c = \frac{\beta_0}{B + B_0} \tag{2.3}$$

Herein β_0 and B_0 are constants. For the application of *RE*BCO tapes and technical superconductors in general it is crucial to know the critical values of the conductor and to provide a reasonable safety margin during normal operation.

2.1.2 Structure of *RE*BCO conductors

The structure of *RE*BCO conductors is a result of the manufacturing process and the need for electrical, thermal and mechanical stabilization. A *RE*BCO based conductor consists of several layers of different materials on top of each other, wherein each layer or material fulfills one specific need [RLS13], [HXS10].

The base material of the conductor is stainless steel or a nickel compound used as a substrate. On this substrate a buffer layer and then the superconductor material is applied. A silver layer directly on the superconducting layer provides mechanical protection, electrical and thermal stabilization and serves as diffusion barrier during the manufacturing process. Thermal and electrical stability can be further improved and tweaked by a coating copper layer. Figure 2.5 shows this typical layered structure of a *RE*BCO tape.

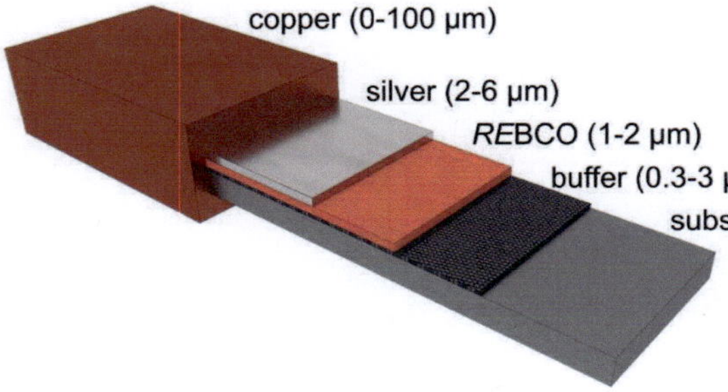

Figure 2.5: Typical layered structure of a *RE*BCO conductor. Thickness and width of the layers are not to scale

The substrate is typically made of nickel-wolfram alloys or nickel-chrome-molybdenum, between 50 µm and 100 µm thick and provides the necessary mechanical stability and flexibility of the conductor, since the *RE*BCO itself is rather brittle [DSH10], [SDG14]. Ideally the substrate has a high resistance and is non-magnetic in order to minimize AC-losses [AMS00], [DGL05]. The buffer layer separates the *RE*BCO from the substrate. It is responsible for the correct orientation of the *RE*BCO lattice and prevents contamination of the superconducting layer e.g. due to diffusion processes from the substrate. The orientation of the lattice must be maintained along the length of the tape in order to enable the superconductivity of the *RE*BCO layer. This is achieved by texturing the surface of the buffer layer or the surface of the substrate directly. Different manufacturing methods for commercially available *RE*BCO tapes are established. In Ion beam assisted deposition (IBAD) and inclined substrate deposition (ISD) processes the texturing is implemented in the buffer layer, while the rolling assisted biaxial textured substrate (RaBiTS) process implements the texture within the substrate [Scha09]. Depending on the manufacturing process the *RE*BCO tape parameters and performance may vary in terms of mechanical stability, magnetization, AC-losses, and throughput [GAK01], [XKZ09], [XSM09], [RLT09], [SCK11], [ZLF14]. The thickness of the buffer layer varies between 0.2 µm (IBAD and RaBiTS) and 3 µm (ISD).

The *RE*BCO elementary cell has an orthorhombic perovskite structure, wherein the CuO planes are the decisive structural factor for the superconductivity [LGFP01]. Due to the texturing the elementary cells are aligned in such a way, that these CuO planes are in parallel to the substrate. Hence, the *RE*BCO tape shows an anisotropic behavior with stronger superconductivity parallel to the *RE*BCO tape than in perpendicular direction. During the manufacturing process the *RE*BCO elementary cells form a polycrystalline structure on the substrate with grain boundaries. The critical current I_c of the *RE*BCO tapes is dependent on the orientation of these grain boundaries [DCML88]. For reasonable lengths of *RE*BCO conductors angles of 6° to 7° have been reported [SCX09]. Doping of the *RE*BCO, for example with zirconium oxide, allows improving the properties of the tape in terms of reduced dependency of the critical current I_c on the angle of an external magnetic field [SCK11]. The superconducting layer has a typical thickness of 1 µm to 2 µm [SCX09].

The superconducting layer is in general mechanically protected and electrically and thermally stabilized by a silver layer (or a gold layer in special cases), which is applied by e.g. sputtering. The silver layer ranges from a typical thickness of 2 µm up to 6 µm. Copper, Brass and stainless steel are suitable materials in order to provide additional electrical, thermal and mechanical stabilization.

For the application in a superconducting fault current limiter the determination of the thickness of the electrical and thermal stabilization is a vital part [APY07]. During fault limitation the current transits from the superconducting layer to the stabilization layer(s). Due to the resistance of the stabilization ohmic losses occur and therefore joule heat is generated. A good thermal stabilization is important in order to quickly dissipate the heat before hot spots are developed and eventually cause the burn-out the *RE*BCO tape. The thermal stability can be improved by increasing the thickness of the copper and silver stabilization layer. On the contrary a high resistance, and therefore a low thickness of the copper and silver layer, is beneficial for the current limitation of a fault current. Consequently the determination of the thickness of both stabilization layers is mostly a compromise between sufficient thermal stability and sufficient resistance to effectively limit a fault current.

2.1.3 State-of-the-art *RE*BCO conductors

As described in chapter 2.1.2 the *RE*BCO crystals must be aligned in a proper manner to enable the superconductivity. Furthermore *RE*BCO is rather brittle and requires mechanical, thermal and electrical stabilization. These constraints make the manufacturing of *RE*BCO tapes a challenging task. Commercially available *RE*BCO tapes differ in several aspects, such as geometry, maximum and homogeneity of critical current I_c and stabilization. This allows a certain degree of customization towards the application. Nowadays, *RE*BCO conductors are ready for applications, but the *RE*BCO conductor must be evaluated carefully in order to decide if it meets the requirements.
*RE*BCO conductors are available from several commercial suppliers, among them American Superconductors (AMSC), Bruker EST, Fujikura, SuNAM, SuperOX,

Superpower and Theva. The technical data provided by selected manufacturers is summarized in Table 2.1.

A few suppliers are capable of manufacturing pieces of up to couple of hundred meters. The price of a REBCO tape has dropped in the recent years from > 300 $/(kAm) [Sel10] to ~150 $/(kAm) [Moo14].

A few companies claim to ramp up their production and manufacture a single piece length of ≥ 1000 m in the next two years and ≥ 2000 m in the next five years. In the same time the critical current is expected to increase to $I_c \geq 700$ A at 77 K in self-field for *RE*BCO tapes with 12 mm width. The price however is predicted to decrease and meet the 50 $/(kAm) for market entrance [Moo14], [Haz14].

It is expected, that other suppliers will have no other option than follow this trend and try to optimize their production process towards increasing piece length and better overall tape performance.

Table 2.1: Provided data of commercial available *RE*BCO tapes of selected manufacturers

Manufacturer	product name	width / mm	critical current I_c (77 K / self-field)	tensile stress (RT)	substrate	stabilizer	manufacturing process
American Superconductor (AMSC)[a]	Amperium 8700	4.24 - 4.55	Standard: 70 A - 100 A Beta: 140 A - 180 A	200 Mpa	n.s.	Brass	Rabits
	Amperium 8501	4.70 - 4.95	80 A - 100 A	150 Mpa	n.s.	Copper	
	Amperium 8502	11.9 - 12.3	250 A - 350 A	150 Mpa	n.s.		
	Amperium 8612	11.9 - 12.3	400 A - ≥ 500 A	200 Mpa	n.s.	Stainless steel (2*75 µm)	
	Amperium 8602	11.9 - 12.3	200 A - ≥ 250 A	200 MPa	n.s.		
Bruker EST[b]		4 or 12	135 A (4 mm) or 420 A (12 mm)	650 Mpa	stainless steel & Hastelloy© C276 50 µm or 100 µm	Silver, gold or copper 0.1 µm to 40 µm	PLD
Fujikura[c]	FYSC-SC05/-SC10 FYSC-S05/-S10	5 (SC05/S05) 10 (SC10/S10)	> 250 A (SC05/S05) > 500 A (SC10/S10)	n.s.	Hastelloy© (75 µm) (SC05,S05, SC10, S10)	Copper (75 µm) (SC05, SC10)	IBAD, PLD
		5 (SC05/S05) 10 (SC10/S10)	> 250 A (SC05/S05) > 500 A (SC10/S10)	n.s.	Hastelloy© (100 µm) (SC05,S05, SC10, S10)	Copper (100 µm) (SC05, SC10)	
Sunam[d]	AN, CN, LB/LS	4 or 12	Standard: 100 A, 150 A, 200 A (4 mm) 400 A, 500 A, 600 A (12 mm) Special: > 200 A, < 100 A (4 mm) > 600 A, < 400 A (12 mm)	> 250 Mpa	Hastelloy© (60 ± 3 µm) non-magnetig stainless steel (105 ± 4 µm)	AN: Silver (1-1.5 µm) CN: Copper (40 µm) LB/LS: Brass or stainless steel (90 µm)	IBAD
SuperOX[e]	2G HTS tape	4 or 12	> 100 A (4 mm) > 250 A (12 mm)	n.s.	Hastelloy© 276, non-magnetic	Silver (up to 10 µm) Copper (up to 50µm)	
SuperPower[f]	SF*	2 to 12	50 A to 300 A (depending on width)	n.s.	Hastelloy© (50 µm)	Silver (2 µm)	IBAD, PLD
	SCS*	2 to 12	50 A to 300 A (depending on width)	n.s.	Hastelloy© (50 µm)	Silver (2 µm) Copper (40 µm)	

a) [AMS15]; b) [Bru15]; c) [Fuj15]; d) [SuN15]; e) [SOX15]; f) [SP15]

2.2 Conventional current limitation

In today's electrical power grids several conventional measures have been implemented or are subject to research and development in order to effectively and reliably limit fault currents. Figure 2.6 shows an overview of possible measures to effectively limit fault currents adopted from [CIG12].

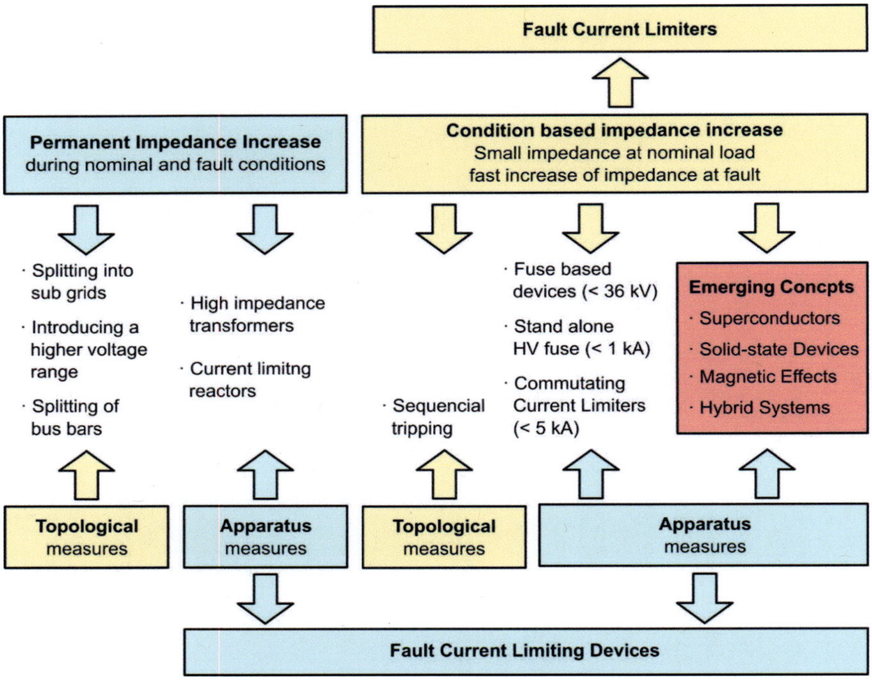

Figure 2.6: Overview of fault current measures according to [CIG12]

The major non-superconducting devices to limit fault currents are discussed hereafter, the major types of superconducting fault current limiters are discussed in chapter 2.3.

Fuses

Fuses are a common, fast and cost-effective measure to limit the first peak of fault currents for low and medium voltages (< 36 kV), which limit the fault current using a melting conductor embedded in quartz sand. During normal operation the fuse has a negligible impedance. In case of a fault current, which is substantially higher than the rated current, the metal inside the fuse is heated and the resistance of the conductor changes rapidly. Within a finite time t_{melt} [AMS15] the heat generated in this process melts the conductor and extinguishes the fault current within a finite time t_{ext} with a lightning arc. This melting process is completed within a fourth half-cycle, essentially preventing the fault current from reaching its unlimited peak. Fuses are designed in such a way, that the following extinguishing process takes usually around t_{ext} < 10 ms [Heu07a].

The major disadvantage of fuses is their one time use. After a fault current is limited the fuse must be replaced manually. Utilities must intervene and the time needed for replacement can be costly due to e.g. interrupted industrial processes. Currents, which are not substantially higher than the rated current may trigger a slower release process. This time-current dependent behavior allows short-termed currents higher than the rated current like inrush currents for example, but must be taken into account when applying fuses. Furthermore, fuses cannot be applied, if fault currents can occur, which exceed the breaking capacity of the fuse. In this case the lightning arc might not be extinguished automatically and the fault current is not interrupted properly. This generally prevents the use of fuses in high voltage grids.

Solid state breakers

Solid state breakers use high power semiconductors to limit fault currents. Possible switching devices for this kind of fault current limiter are gate turn-off (GTOs) thyristors, integrated gate commutated thyristors (IGCTs) and insulated gate bipolar transistors (IGBTs).The fault current limitation is achieved by actively triggering the switching devices and commutate the fault current into a dedicated current-limiting branch or dedicated energy absorbing device [KSD12]. A variety of circuit topologies are suitable to limit fault currents and are under research and investigation [AS12]. An example of such a circuit or topology is shown in Figure 2.7 using a GTO [MSD04].

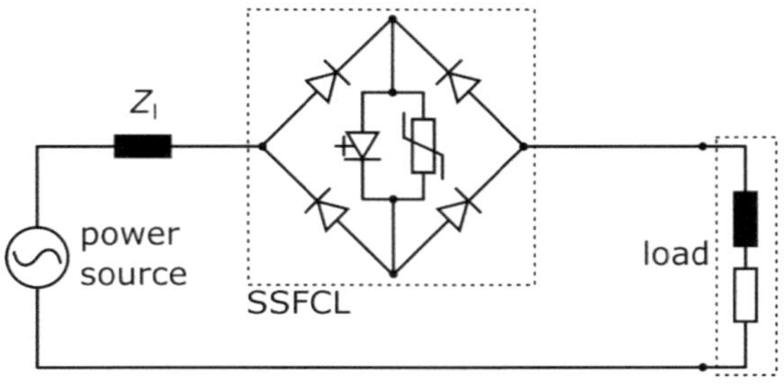

Figure 2.7: Topology of an SSFCL based on a GTO and diode bridge rectifier circuit. The GTO is placed in the DC branch parallel to a varistor [MSD04]

The current limiting circuit is placed in the DC branch of a diode rectifier bridge circuit, which contains the GTO and a varistor connected in parallel. During normal operation the AC current is conducted through the circuit, and the rectified DC current flows unidirectional through the GTO. If a fault occurs the GTO is turned off and the current is commutated into the varistor branch, which limits the current.
On-state losses occur during normal operation. The losses and costs increase with voltage and current, since for higher voltages more semiconductor devices must be connected in series until the cumulated blocking voltage of the devices exceeds the grid voltage.

Air core reactors

Basically air core reactors (ACR) are a solenoidal coil with a winding made of conventional conductor like copper or aluminum. Usually the air core reactor is designed in such a way, that the resistance of the winding is much lower compared to the reactance. Nevertheless, ohmic losses occur in in the winding and heat the conductor. Therefore, the surface of the winding must be large enough to allow a sufficient heat transfer to the surrounding air. Figure 2.8 (A) shows the typical structure of an air core reactor. The winding is divided into sections separated by an air gap to increase the surface and improve the cooling. Within the winding sections the current carrying cross-section of the conductor is separated in several wires to facilitate the winding process by lowering the necessary bending force.

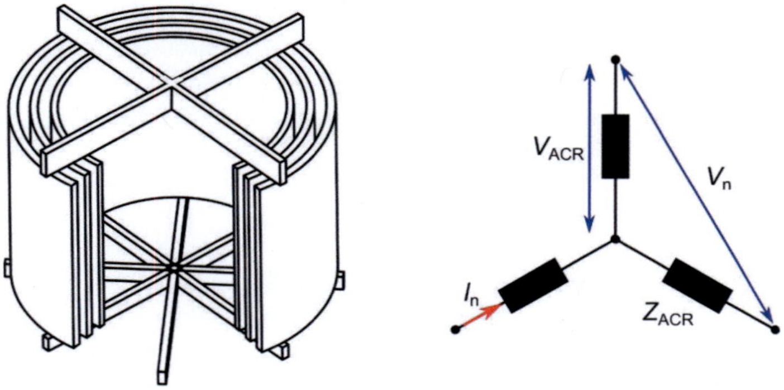

(A) (B)

Figure 2.8: (A) Schematic of an air core reactor and (B) equivalent circuit with annotations for 3 air core reactors in Y-connection

By variation of the geometry air core reactors can be designed for any apparent power, voltage level and impedance. The rated three phase power of the ACR can be calculated by

$$S_{\mathrm{ACR}} = \sqrt{3} \cdot V_{\mathrm{n}} \cdot I_{\mathrm{n}} \tag{2.4}$$

Herein V_n is the nominal line voltage, I_n the nominal current conducted through the ACR (compare Figure 2.8 (B)). The relative voltage drop v_{ACR} across the air core reactor can be calculated by

$$v_{\mathrm{ACR}} = \frac{V_{\mathrm{ACR}}}{V_{\mathrm{n}}/\sqrt{3}} = \frac{Z_{\mathrm{ACR}} \cdot I_{\mathrm{n}}}{U_{\mathrm{n}}/\sqrt{3}} \tag{2.5}$$

Wherein Z_{ACR} is the impedance of the air core reactor. Assuming, that the air core reactor is mainly inductive the impedance can be replaced by the reactance in eq. (2.5) and the insertion of eq. (2.4) gives

$$v_{\mathrm{ACR}} = \frac{X_{\mathrm{ACR}} \cdot S_{\mathrm{ACR}}}{V_{\mathrm{n}}^2} \tag{2.6}$$

Herein the reference voltage drop v_{ACR} is equal to the reference impedance z_{ACR}. Furthermore, if the resistance of the winding is negligible compared to the reactance the voltage drop can be assumed as purely inductive and thus only determined by geometrical parameters.

The impedance of the air core reactor, which is necessary for effective fault current limitation, however, applies during normal or load operation as well. The resulting voltage drop during normal operation decreases the system stability, the maximum power transmission and generally cause system perturbations. For this reason the reference impedance of commercial air core reactors is usually in the range from 3% to 10%.

2.3 State-of-the-art superconducting fault current limiters

The electrical properties of superconductors, namely the transition from superconducting to normal state, can be utilized in several ways to effectively limit a fault current. However, the operational behavior of all different types of SFCLs is the same as shown in Figure 2.9.

During normal operation the transport current is not affected by the SFCL. As soon as a fault occurs the current is rising. If no measures are taken to limit the fault current reaches the prospective current i_p in the first half cycle. Using a fault current limiter the prospective current i_p is limited, effectively reaching the limited current peak i_{lim}, which is significantly lower than i_p. The ratio of i_p/i_{lim} depends on the type of the fault current limiter and design implementation. The unlimited fault continues until the fault is cleared in a controlled manner by opening a breaker. In any case, the load flow is interrupted.

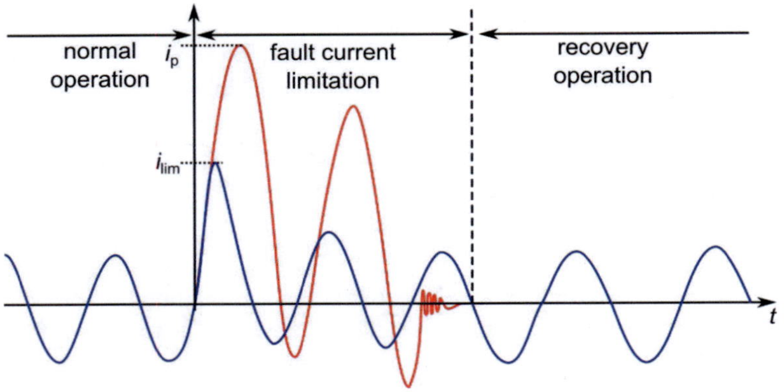

Figure 2.9: Operational modes of a superconducting fault current limiter. The red curve shows the unlimited fault current and the blue curve the limited fault current

A SFCL is capable of controlling the fault current through the whole duration of the fault and continue with normal load after the fault. This operation is called recovery under load. The recovery time mainly depends on the type of SFCL, design implementation, fault duration and prospective current.

Resistive type SFCL

The resistive SFCL directly uses the transition of a superconductor from superconducting state to normal conducting state, if a current is higher than the critical current I_c. The principle of the resistive type SFCL is shown in Figure 2.10.

During normal operation the load current is carried exclusively by the superconductor in superconducting state with no measureable resistance (R_{sc} = 0). If a fault current occurs the superconductor transits to normal state and develops a high resistance R_{sc}. As a result, the current is commutated to a parallel resistance R_{par}. In case of *RE*BCO tapes R_{par} is determined by the thickness of the silver and a possible copper layer or an external shunt. To protect the complete *RE*BCO tape an additional impedance Z_s can be connected in parallel, such as an air core reactor for example [NSK09].

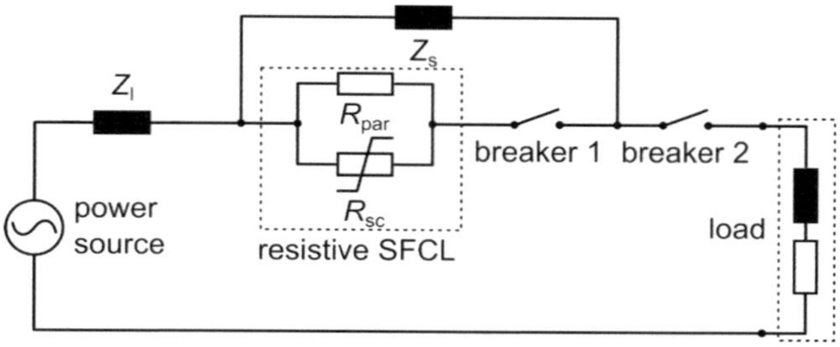

Figure 2.10: Circuit diagram of a resistive type SFCL with parallel impedance Z_s and circuit breakers

Resistive type SFLCs can be built relatively compact and have negligible impedance during normal operation. The major drawback of the resistive type SFCL is, that they require current leads from room temperature to the operating temperature of the superconductor (77 K for LN_2 bath). This causes substantial losses through the current leads during normal operation mode. Due to the quenching of the superconductor the recovery time can last up to several seconds [SKN07], [BNK11].

Resistive type SFCLs can currently be considered the most mature type for medium voltage applications. Feasibility research and conceptual designs have been reported even before the discovery of HTS [GF78]. The research and development has been continued ever since resulting in wide field testing [BEB05], [BBW05], [NSK09], [HYY11], [EKB12], [MBA13], [MBA15] and commercial availability nowadays [DKH10], [BHK11], [BBD11], [BHS15].

DC biased iron core type SFCL

The DC biased iron core SFCL uses an arrangement of a superconducting coil and two normal conducting coils coupled with iron cores as shown in Figure 2.11 (A). The load current I_{AC} flows through the normal conducting coils, which are connected to the power system. A DC-current I_{DC} in the superconducting coil saturates both iron cores. Hereby the current I_{DC} must be high enough to maintain the saturated state during normal operation. Hence the magnetic flux density is only oscillating in a small region of the B-H curve as indicated in Figure 2.11 (B). In this region the permeability μ_r is approximately that of air and therefore the impedance of the AC coils is comparable to that of air core reactors. During a fault the current rises and de-saturates the iron core, moving to a region with a high permeability μ_r. The impedance increases accordingly and limits the fault current. [MD09].

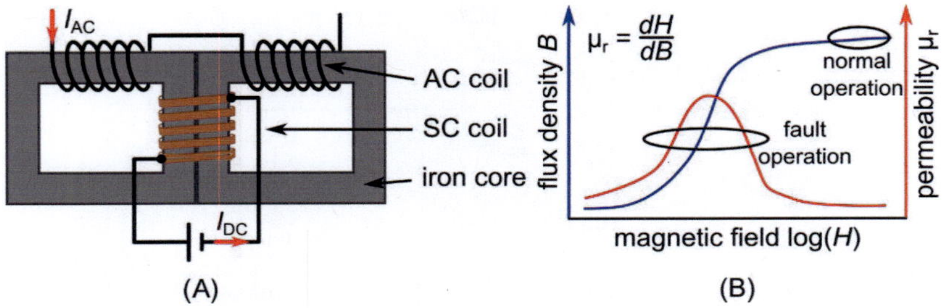

Figure 2.11: (A) Schematic of the saturated iron core SFCL for single phase and (B) qualitative magnetization curve with regions of operation according to [MD09]

The saturated iron core SFCL offers almost immediate recovery, because the superconductor is not quenched during fault the. The high current carrying capability of superconductors and no AC-losses allows to keep the needed conductor material and cooling power low. However the need for an iron core makes this kind of SFCL relatively heavy and induced currents in the superconducting coil during fault operation must be suppressed.

The concept of the DC biased iron core SFCL was first successfully tested in 1982 by a 3 kV, 550 A working prototype, which proved the concept [RPB82]. This original design could be improved in terms of size and protection of the superconducting coil [HDB05], [RFW07], [MD09]. First saturated iron core SFCL have been reported to be in the stage of field testing [XHW11], [MRD11], [XGS13].

Shielded iron core SFCL

Basically the shielded iron core SFCL is an inductive type SFCL. The load current is conducted in a conventional solenoid winding. A superconducting winding is coupled to the conventional winding via an iron core as shown in Figure 2.12. During normal operation the induced current is lower than the critical current I_c of the superconductor. This is effectively shielding the iron core, compensating the magnetic flux within the iron core. Hence the impedance of the iron core SFCL during normal operation is only determined by the leakage flux between the windings. In case of a fault current the induced current quenches the superconductor and a resistance is generated as it is the case for the resistive type SFCL. The shielding collapses and the fault current is limited by an impedance generated by the magnetized inductance of the iron core and the resistance of the superconducting winding in normal state.

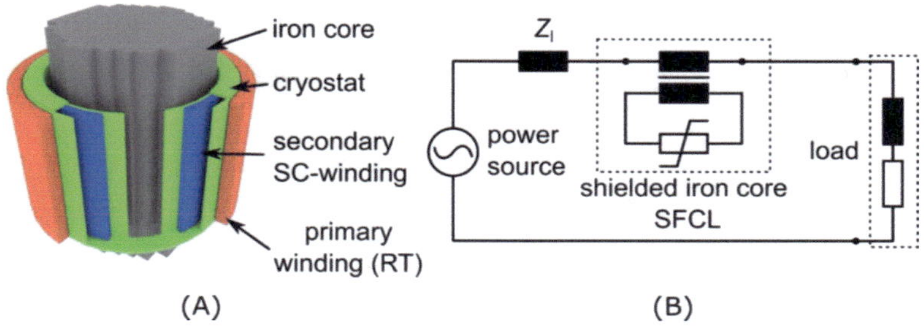

Figure 2.12: (A) Schematic of the shielded iron core SFCL and (B) equivalent circuit diagram

An advantage of this concept is the absence of current leads to low temperatures, which reduces the cryogenic losses compared to the resistive type SFCL. By adjusting the number of turns of both windings the superconductor in the secondary winding can be exposed to lower voltages and higher currents, which is beneficial for the use of superconductors. The use of an iron core makes this type of SFCL rather heavy and in size comparable to a transformer of the same power rating.

The investigation of the shielded iron core SFCL started in the early 1990s [BFP91], [FBA93]. Working prototypes and demonstrators have been reported in the same decade [MSG95], [IO95], [CWN97], [KI97] and one SFCL has been field-tested for one year [PLR97]. The research on this SFCL type has continued since then, but remains within theoretical investigations, lab-scale experiments and models [SMV04], [KJK05], [UMD09], [SMC10], [WJK14].

Coreless inductive SFCLs

Coreless inductive SFCLs (or sometimes referred to as transformer-type SFCLs) are composed of two concentric aligned solenoid windings. One winding, the primary winding, is carrying the load current, while the secondary winding is short-circuited and shielding the primary winding. In principle the functionality and the structure is similar to the shielded iron core SFCL. However the concepts, which have been subject to research and development differ in several aspects. The absence of an iron core lowers the coupling between the windings and the impedance during fault. To increase the impedance during fault it is beneficial, if the windings generate a high resistance. To increase the coupling, the flux leakage is lowered by minimizing the distance between the windings. Usually this is achieved by operating both windings within one cryostat at cryogenic temperatures. Hence this type of SFCL needs current leads and suffers from the same losses as the resistive type SFCL.

Early concepts based on low temperature superconductors intended to adjust the current at which the SFCL would effectively start limiting a fault current [FSN99], [SFH99], [HNC04]. By sliding the secondary winding in axial direction the magnetic coupling of the solenoids is altered and therefore it is possible to adjust the trigger level at which a fault current is effectively limited. The generated impedance during fault turned out rather low using BSCCO wire for the windings [FNB07], [SBN08]. In order to improve the impedance special winding concepts of the coils have been proposed, implemented and tested with a model SFCLs [NSS09], [SNO09], [ONN11], [SNY12] [SNY13]. However the use of *RE*BCO tapes is generally preferable and superior for this kind of SFCL [JKK07], [YYS15]. Concepts based on *RE*BCO conductors have been proposed and a successful test of a 15 kV SFCL has been reported [KJW10], [KMJ11], [KMK12], [KMK13], [MKK15].

Summary

Table 2.2 summarizes field-tested SFCLs and SFCLs installed in the power grid in the last years. Until now the resistive type and the DC biased iron core SFCL proved their capability for operation within power systems.

Table 2.2: Overview of major SFCL field tests

Year	Country	Type	Lead Company	Data	Ref.
2004	Germany	Resistive	ACCEL / Nexans SC	12 kV, 600 A	[BEB05]
2005	China	Diode-Bridge	CAS	10.5 kV, 1.5 kA	[LGX07]
2007	Korea	Resistive	KEPRI	22.9 kV, 630 A	[HYY11]
2008	Japan	Resistive	Toshiba	6.6 kV, 72 A	[YKM09]
2008	China	DC biased iron core	Innopower	35 kV, 90 MVA	[XHW11]
2009	UK	Resistive	Nexans SC	12 kV, 100 A	[DKH10]
2009	Germany	Resistive	Nexans SC	12 kV, 800 A	[BBD11]
2010	USA	DC biased iron core	Zenergy	12 kV, 1.25 kA	[MRD11]
2011	Germany	Resistive	Nexans SC	12 kV, 800 A	[EKB12]
2012	Italy	Resistive	RSE	9 kV, 220 A	[MBA13]
2012	China	DC biased iron core	Innopower	220 kV, 300 MVA	[XGS13]
2012	UK	Resistive	Nexans SC	12 kV, 400 A	[BHK11]
2013	Germany	Resistive	Nexans SC	12 kV, 2.3 kA	[EKB12]
2015	Italy	Resistive	RSE	9 kV, 1 kA	[MBA15]

Figure 2.13 shows the field-tested SFCLs and SFCLs installed in the power grid by voltage and current rating. It can be seen, that the resistive type SFCL is preferably used for medium voltage.

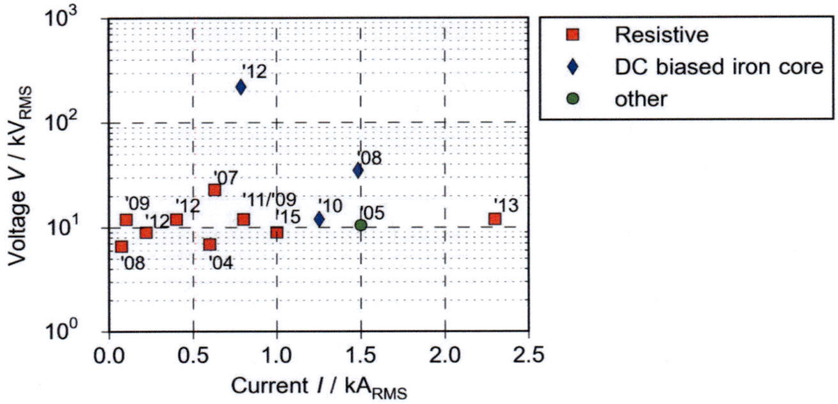

Figure 2.13: Rated voltage and current of selected field tested SFCLs (compare Table 2.2)

2.4 Air Coil Superconducting Fault Current Limiter

Generally the air core superconducting fault current limiter (AC-SFCL) can be classified as a coreless inductive SFCL. Like this type of SFCL it consist of two concentric aligned solenoid windings, which are inductively coupled. However, the fundamental concept of the AC-SFCL is unique and differs from other coreless concepts. The initial objective is to improve the air core reactor described in chapter 2.2 by retrofitting with a superconducting insert using stacked *RE*BCO rings as shown in Figure 2.14. The objective is to minimize the inductance of the air core reactor in normal operation, while maintaining the same level fault current limitation during fault operation. For this retrofit the construction of the air core reactor remains unaltered and assumes the function of a primary winding. This implies that the primary winding is operated at ambient temperature and only the superconducting winding of the insert is cooled down to cryogenic temperatures (77 K). Therefore the electrical and thermal insulation of the superconducting winding must be placed between both windings as shown in Figure 2.14 (B). The superconducting winding assumes the function of a secondary, short-circuited winding.

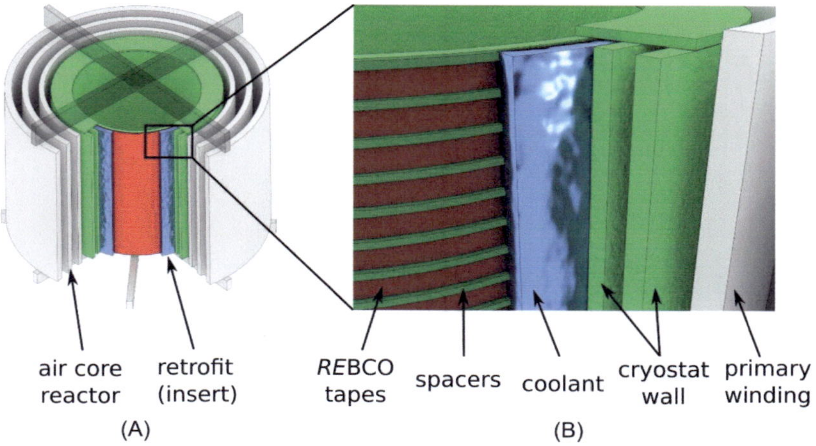

Figure 2.14: Schematic of the AC-SFCL: (A) Air core reactor with insert and (B) magnification of the insert showing the insulation and the winding

In operation any alternating current in the primary winding induces a current in the secondary winding. Any material used for the cryostat and insulation between the windings should be non-conductive material, such as G10 to prevent additional magnetic coupling and disturbances. The magnetic field generated by the induced current in the secondary winding counters the magnetic field of the primary winding in the shared air core of both windings. During normal operation the induced current is lower than the critical current I_c of the superconductor and the magnetic field in the air core is fully compensated and only a much smaller magnetic stray field remains between the windings. The magnetic field distribution in case of the AC-SFCL during normal operation compared to the air core reactor is shown in Figure 2.15.

The displacement of the magnetic field, due to the shielding minimizes the inductance and hence the impedance of the AC-SFCL during normal operation compared to the air core reactor. Subsequently the ratio of impedance during fault and during normal operation is determined by the ratio of magnetic field during fault and magnetic stray field during normal operation. The calculation of the distribution of the magnetic field between the windings and the effect on the inductance is discussed in chapter 3.2. Furthermore the geometry of the primary winding can be optimized in order to minimize the stray field and therefore the impedance during normal operation z_n.

If a fault current occurs the induced current exceeds the critical current I_c of the superconductor. The superconductor quenches, transits to normal state and generates a resistance, which limits the induced current. Consequently the shielding collapses and an additional impedance is generated, which is significantly higher than in normal operation effectively limits the fault current.

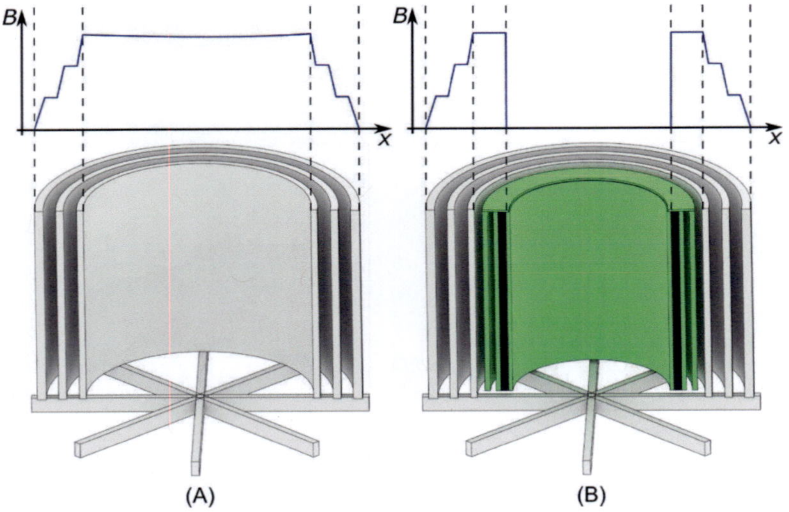

Figure 2.15: Magnetic field distribution in case of (A) the air core reactor and (B) the AC-SFCL during normal operation

In order to enable the operation as described above, several aspects and parameters of the secondary winding must be controlled and adjusted carefully to enable an effective shielding of the primary winding during normal operation and at the same time an effective fault current limitation.

The secondary winding is assembled using single, short-circuited superconducting *RE*BCO rings, which are stacked on top of each other as shown in Figure 2.16.

To prevent movement of the *RE*BCO tapes due to the magnetic forces during fault limitation the *RE*BCO tapes must be separated and positioned axially with spacers as indicated in Figure 2.16 (B). At the same time the radial contraction of the *RE*BCO rings due to the thermal cool down must be ensured.

The stacking of short-circuited rings offers several advantages in comparison to a solenoidal winding:

- The reduced piece length allows selecting *RE*BCO tape pieces with homogeneous distribution of critical current I_c, which lowers the probability of hot-spots.
- Depending on the position in the winding the perpendicular and parallel magnetic field components vary. *RE*BCO tapes with different behavior in magnetic field can be selected and placed optimally.
- The critical current I_c of each *RE*BCO ring can be selected individually in order to influence the quench behavior of the winding.

- The electrical and thermal stabilization of each ring can be adjusted individually
- A failure, e.g. burn-out, of a single ring does not affect the operation of the other rings and minor effects the operation of the complete AC-SFCL.
- A defective *RE*BCO ring can be replaced easier than a complete winding.

The space between the *RE*BCO rings or the thickness of the spacers respectively, must be minimized for optimal shielding and low impedance during normal operation.

(A) (B)

Figure 2.16: (A) Conventional solenoid winding and (B) winding consisting of stacked rings with spacers

Beyond the retrofit application the concept of the AC-SFCL can be generalized as a stand-alone SFCL. In this case the air core reactor or primary winding respectively can be redesigned and optimized for the superconducting insert. The impedance of the optimized primary winding remains the same as for the conventional air core reactor. However optimizing the geometry of the primary winding allows improving the electrical parameters of the AC-SFCL. Using an optimized primary winding the magnetic stray field is minimized and therefore the impedance during normal operation as well. The electrical parameters strongly depend on the geometry. This dependency and the behavior during normal operation and during fault is described in chapter 3.1.

Table 2.3 shows the advantages and disadvantages of the described devices for current limitation and compares them with the AC-SFCL.

The AC-SFCL shows the same advantages as other types of SFCLs (self-triggering, self-recovering). Compared to inductive type SFCLs, such as DC biased iron core and shielded iron core, the AC-SFCL has a significant weight advantage due to the absence of the heavy iron core. Compared to the resistive type SFCL the AC-SFCL has reduced losses due to the absence of current leads. However the impedance during normal operation is not as low as it is for the resistive type SFCL, but is significantly reduced compared to the air core reactor and should therefore have minimal effects on the power system.

Table 2.3: Comparison of Advantages and Disadvantages of current limiting devices

Current limiting solution	Advantages	Disadvantages
Fuses	• Low cost	• One time use • Needs replacement • Limited scalability
Solid-state devices	• Fast switch-off of fault (µs)	• Losses scale with voltage and current as well as demand for semiconductor switches
Air core reactors	• Easy to install	• Adds impedance during normal operation • Voltage drop • May cause instabilities
Resistive type SFCL	• Negligible impedance under load • Self-triggering • Self-recovering • Compact	• Require current leads, which increases losses
DC biased iron core SFCL	• No quenching of superconductor • Self-triggering • Self-recovering	• Weight and size • Only fail-safe with uninterruptible power source
Shielded iron core SFCL	• No current leads • Self-triggering • Self-recovering	• Weight (iron core)
Air Coil SFCL	• No current leads • No iron core • Self-triggering • Self-recovering • Low impedance during normal operation	• Impedances depending on size of primary winding

3 Design method for an Air Coil Superconducting Fault Current Limiter

3.1 Equivalent circuit diagram and operation modes

The Air Coil Superconducting Fault Current Limiter consists of two concentric aligned solenoids with the same height as shown in Figure 3.1.

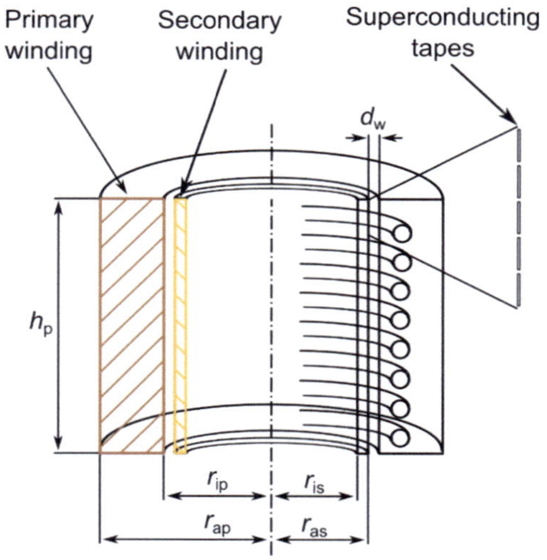

Figure 3.1: Schematic cross-sectional view of the AC-SFCL

The primary coil or primary winding carries the normal current I_n. This winding is made of copper or aluminum wire and operates at room temperature. Typically the turns are distributed rather sparse over the cross-section of the winding (fill-factor f_{fp} < 0.7) in order to provide sufficient cooling.

Primary and secondary winding have a distance d_w for a sufficient thermal and electrical insulation. This distance d_w determines the outer diameter of the secondary winding.

The secondary winding consists of superconducting *RE*BCO tapes, which are aligned upright with a small gap between the tapes in order to shield the primary winding during nominal operation. Ideally, one layer of *RE*BCO tapes is sufficient to carry the induced current during normal operation.

The distance between the primary and secondary winding affects the performance, the critical parameters and impedance during nominal operation. With increasing winding distance d_w, the magnitude of the magnetic field caused by the current in the primary winding decreases. Hence the induced current in the secondary winding decreases as well and *RE*BCO tapes with respective minimum critical current $I_c(B,T)$ can be used. The winding distance d_w determines the stray field across the windings and influences the impedance z_n during normal operation (compare chapter 3.2).

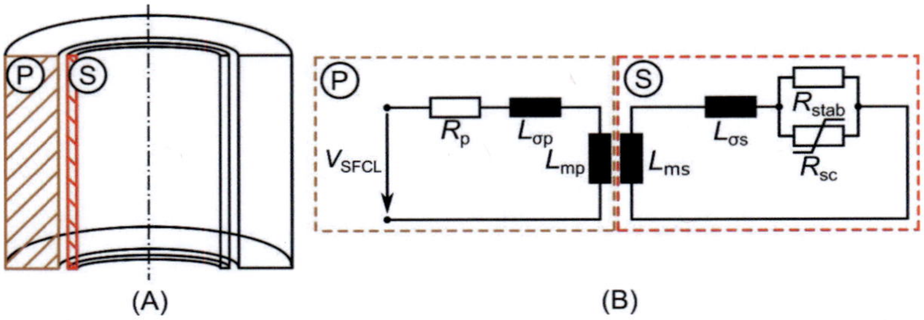

Figure 3.2: (A) Cross-section of the AC-SFCL and (B) equivalent circuit diagram showing the respective circuit for primary and secondary winding.

Figure 3.2 shows the cross-sections of the primary winding ("P") with the secondary winding ("S") and the corresponding circuit diagrams representing the windings. Herein R_P is the primary resistance, $L_{\sigma p}$ the primary stray inductance and L_{mp} the main inductance of the primary winding. Accordingly L_{ms} is the main inductance of the secondary winding and $L_{\sigma s}$ the stray inductance of the secondary winding. The non-linear behavior of the superconductor is represented by the parallel circuit of the resistance R_{sc} and the resistance R_{stab}. The resistance R_{stab} represents the resistance of the electrical stabilization of the superconducting tape. The resistance R_{sc} represents the resistance of the superconducting layer of the *RE*BCO tapes, which is negligible in superconducting state during nominal operation and significantly higher than R_{stab} in current limiting mode.

Since both windings share the same magnetic field the main inductances L_{mp} and L_{ms} can be combined into one main inductance L_m. The parameters of the secondary winding can be transformed on the primary side using the following prescriptions:

$$L'_{\sigma s} = c_f \cdot L_{\sigma s} \tag{3.1}$$

$$R'_{stab} = c_f \cdot R_{stab} \tag{3.2}$$

$$R'_{\mathrm{sc}} = c_{\mathrm{f}} \cdot R_{\mathrm{sc}} \tag{3.3}$$

Herein is c_f the coupling factor. Ideally c_f is defined by the ratio of the number of turns of the primary winding N_p and the number of turns of the secondary winding N_s [Schw06].

$$c_{\mathrm{f}} = \left(\frac{N_{\mathrm{p}}}{N_{\mathrm{s}}}\right)^2 \tag{3.4}$$

If the coupling is not ideally, it is determined by the ratio of the main inductances [Phi00]:

$$c_{\mathrm{f}} = \frac{L_{\mathrm{mp}}}{L_{\mathrm{ms}}} \tag{3.5}$$

The equivalent circuit diagram incorporating these transformations is shown in Figure 3.3 for the two modes of operation of the AC-SFCL.

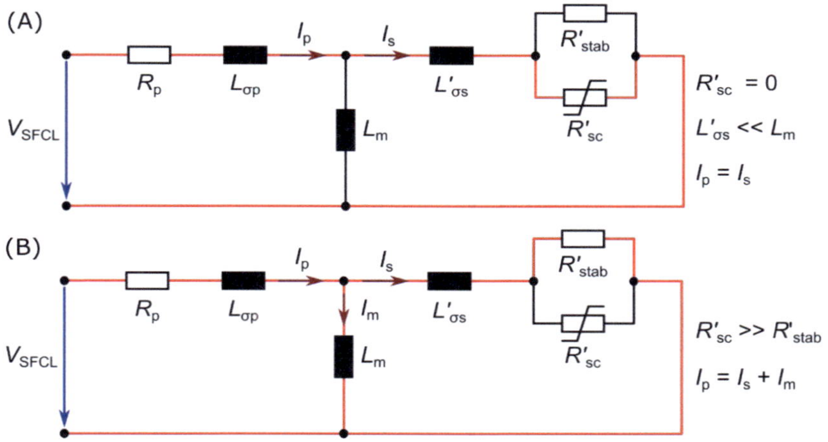

Figure 3.3: Modes of operation: (A) During normal operation the main inductance L_m is short-circuited and (B) during fault current limitation the resistance R'_{stab} and inductance L_m contribute to the total impedance

During normal operation the current induced in the secondary winding is lower than the critical current I_c of the superconducting *RE*BCO tape (compare Figure 3.3 (A)). The stray inductance $L'_{\sigma s}$ of the secondary winding is significantly lower than the main inductance L_m and the resistance R'_{sc} is negligible. Practically the main inductance L_m is short-circuited and the total impedance of the circuit can be calculated by:

$$\underline{Z}_n = R_p + j\omega\left(L'_{\sigma s} + L_{\sigma p}\right) \tag{3.6}$$

Herein ω is

$$\omega = 2\pi \cdot f_\mathrm{n} \tag{3.7}$$

with f_n being the nominal frequency (usually 50 Hz or 60 Hz). If a fault occurs the induced current in the secondary winding will quench the superconductor and the secondary current is conducted in the stabilization layer (compare Figure 3.3 (B)). The resistance R'_{stab} is significantly lower than the resistance R'_{sc}. The impedance Z_{lim} of the circuit during fault current can be calculated by:

$$\underline{Z}_{lim} = j\omega L_{\sigma p} + R_p + \frac{(j\omega L'_{\sigma s} + R'_{stab}) \cdot j\omega L_m}{(j\omega L'_{\sigma s} + R'_{stab}) + j\omega L_m} \tag{3.8}$$

Formulas and equations to calculate the resistances and inductances of the equivalent circuit diagram are derived in chapter 3.2.

3.2 Requirements and design equations

Specification

The fundament for the calculation of the AC-SFCL is the electrical specification. These parameters are defined by the operating condition in the power grid. In principle the AC-SFCL can be designed for any medium voltage (10 kV to 40 kV) up to ultra-high voltage (400 kV) for a specific apparent power.

Table 3.1: Electrical specification

Electrical parameter	Symbol	Unit
Apparent power	S_n	VA
Nominal voltage	V_n	Volts
Nominal frequency	f_n	Hz
Impedance during fault operation	z_{lim}	%
Impedance during nominal operation	z_n	%

From the specified values in Table 3.1: the nominal current conducted in the primary winding can be derived assuming a three phase system by

$$I_n = \frac{S_n}{\sqrt{3} \cdot V_n} \tag{3.9}$$

The actual impedance Z_{lim} can be obtained from the following correlation

$$z_{lim} = Z_{lim} \frac{\sqrt{3} \cdot I_n}{V_n} \tag{3.10}$$

In case of the primary winding the impedance Z_{lim} is determined by the resistance R_p and the reactance X_p of the primary winding:

$$Z_{lim} = \sqrt{R_p^2 + X_p^2} \tag{3.11}$$

Herein the resistance R_p can be calculated using eq. (3.76) and the reactance X_p by

$$X_p = 2\pi \cdot f_n \cdot L_p \tag{3.12}$$

With the primary inductance L_p calculated using eq. (3.50). The impedance during normal operation z_n is assumed to be lower than the impedance during fault operation by a reasonable factor f_z

$$z_n = \frac{z_{lim}}{f_z} \tag{3.13}$$

Practically f_z is selected to allow an impedance during normal operation of $z_n \leq 1\%$.

Geometrical parameters

Two different cases must be considered regarding the geometrical parameters. If the AC-SFCL is designed as a retrofit for an air core reactor the geometry is mostly defined. For effective shielding the height of the secondary winding h_s must have at least the same height as the primary winding. Only the diameter of the secondary superconducting winding d_{is} and the distance between the windings d_w respectively are subject to variation.

For an optimized design of the AC-SFCL it is necessary to either define a reasonable value range for the height h_p and inner diameter d_{ip} of the primary winding. The outer diameter d_{ap} and the number of turns N_p are calculated to match the impedance L_{lim} for a given pair of height h_p and inner diameter d_{ip}.

The actual number of turns of the secondary winding N_s does not influence the impedance of the AC-SFCL during current limitation (compare eq. (3.80)). Table 3.2 summarizes all necessary geometrical parameters.

Table 3.2: Geometrical specification of the AC-SFCL

Geometrical Parameter	Symbol	Unit
Height (primary)	h_p	m
Inner diameter (primary)	d_{ip}	m
Outer diameter (primary)	d_{ap}	m
Number of turns (primary)	N_p	-
Winding distance	d_w	m
Fill factor of primary winding	f_{fp}	-
Fill factor of secondary winding	f_{fs}	-

Parameters and constraints: Properties of the superconducting *RE*BCO tape and conventional conductor

The properties of the superconducting tape determine the current limiting behavior and capability. The most significant parameter of *RE*BCO tapes is the critical current $I_c(B,T)$, which defines the maximum current at a certain magnetic field and temperature.

*RE*BCO tapes are commercially available with different widths. Since the tapes will be stacked upright in the secondary winding the width is one parameter to determine the total amount of turns and therefore the total length of *RE*BCO tape required for sufficient shielding.

Each *RE*BCO tape is electrically and thermally stabilized with at least one layer of normal conducting material. Usually silver and copper are used as stabilizer materials. A thin stabilization layer results in a high resistance during fault current limitation. At the same time the thermal and electrical stability is reduced and therefore the risk of burning-out the tape is increased. It is evident, that a good balance between sufficient stabilization and necessary impedance for current limitation is most important for the operation of the AC-SFCL. Table 3.3 summarizes the required parameters of the *RE*BCO tape for the design of an AC-SFCL.

Table 3.3: Major parameters of the superconducting tape

Tape parameter	Symbol	Typical value range
Critical current	I_c	150 A to 600 A (for 12 mm width)
Width	b_{sc}	4 mm to 12 mm
Stabilizer thickness	h_{stab}	2 µm of silver and up to 100 µm of copper

If the AC-SFCL is not designed as a retrofit of an existing air core reactor the primary winding must be designed as well. It consists of conventional conductor, e.g. copper or aluminum. For these materials the resistivity ρ_p (or conductivity σ_p respectively) and the maximum current density J_p must be specified. Since the primary winding is only air cooled at room temperature (RT) the surface of the winding must be large enough to prevent overheating caused by ohmic losses. Practically the primary winding is interspersed with cooling channels, which increase the cross-section. In the design process this is considered by choosing an appropriate fill-factor f_{fp}. Table 3.4 summarizes the required parameters of the *RE*BCO tape for the design of an AC-SFCL.

Table 3.4: Main parameters for the conductor of the primary winding

Parameter	Symbol	Typical value
Resistivity at RT	ρ_p	$0.172 \cdot 10^{-2}$ Ωm/mm² (copper)
Current density	$j_{p,con}$	1 A/mm² to 2 A/mm² (at RT for copper)
Fill factor	f_{fp}	0.2 to 0.7

Calculation of the magnetic field of a solenoid

The calculation of the magnetic field is essential for the design of the AC-SFCL. For the design method an approach based on elliptic integrals was chosen [Pre83], which was already used successfully for design of superconducting transformers [Ber11] and the design of SMES [Nae10]. The basic equations for this approach are outlined hereafter.

For the component-wise calculation of the magnetic field the variables ρ and ζ are introduced, which refer to the inner radius r_i of the solenoid.

$$\rho = \frac{r}{r_i} \tag{3.14}$$

$$\xi = \frac{z}{r_i} \tag{3.15}$$

The basic equation for the calculation of the magnetic field for the radial field component is

$$B_r = \mu_0 \cdot j_e \cdot r_i \cdot b_r(\rho, \xi) \tag{3.16}$$

and the axial field component

$$B_z = \mu_0 \cdot j_e \cdot r_i \cdot b_z(\rho, \xi) \tag{3.17}$$

For both equations a homogeneous distribution of the current density j_e is assumed. The functions in eq. (3.5) and eq. (3.6) are

$$b_r(\rho,\xi) = \alpha\left[f\left(\frac{\rho}{\alpha},\frac{\beta-\xi}{\alpha}\right) - f\left(\frac{\rho}{\alpha},\frac{\beta+\xi}{\alpha}\right)\right] - [f(\rho,\beta-\xi) - f(\rho,\beta+\xi)] \tag{3.18}$$

and

$$b_z(\rho,\xi) = -\alpha\left[g\left(\frac{\rho}{\alpha},\frac{\beta-\xi}{\alpha}\right) - g\left(\frac{\rho}{\alpha},\frac{\beta+\xi}{\alpha}\right)\right] + [g(\rho,\beta-\xi) - g(\rho,\beta+\xi)] + \begin{Bmatrix} \alpha-1 & \rho<1 \\ \alpha-\rho & 1\leq\rho<\alpha \\ 0 & \rho\geq\alpha \end{Bmatrix} \tag{3.19}$$

Herein the functions f and g are elliptic integrals:

$$f(\rho,\xi) = -\frac{1}{2}\int_0^1 \frac{t}{\sqrt{(t+\rho)^2+\xi^2}}\cdot\frac{2}{\pi}\cdot cel[k_c,1,1,-1]\cdot dt \tag{3.20}$$

$$g(\rho,\xi) = -\frac{1}{2}\int_0^1 \frac{\xi}{t+\rho}\frac{t}{\sqrt{(t+\rho)^2+\xi^2}}\cdot\frac{2}{\pi}\cdot cel\left[k_c,\left(\frac{t-\rho}{t+\rho}\right)^2,1,\frac{t-\rho}{t+\rho}\right]\cdot dt + \begin{Bmatrix} \frac{1-\rho}{2} & \rho<1 \\ 0 & \rho>1 \end{Bmatrix} \tag{3.21}$$

wherein

$$cel[k_c,p,a,b] = \int_0^{\frac{\pi}{2}} \frac{a\cdot cos^2\Psi + b\cdot sin^2\Psi}{cos^2\Psi + p\cdot sin^2\Psi}\cdot\frac{1}{\sqrt{cos^2\Psi + k_c^2\cdot sin^2\Psi}}\cdot d\Psi \tag{3.22}$$

and

$$k_c = \sqrt{\frac{(t-p)^2+\xi^2}{(t+p)^2+\xi^2}} \tag{3.23}$$

With the equations above and a defined geometry and current density it is possible to calculate the magnetic field or magnetic field density respectively at any given point in a solenoid and concentric aligned solenoids as well

$$B = \sqrt{B_r^2 + B_z^2} \tag{3.24}$$

Calculation of the stray field between the windings and the stray inductance

Figure 3.4 shows the assumed distribution of the magnetic field across the windings. The slope of the curve shows a linear increase of the magnetic field starting at the inner radius of the secondary winding r_{is} to the outer radius of the secondary winding r_{as}. Between the windings the magnetic field is assumed to be constant at a magnetic field strength H_m. Starting at the inner radius of the primary winding r_{ip} the magnetic field decreases linearly until it deceases at the outer radius of the primary winding r_{ap}.

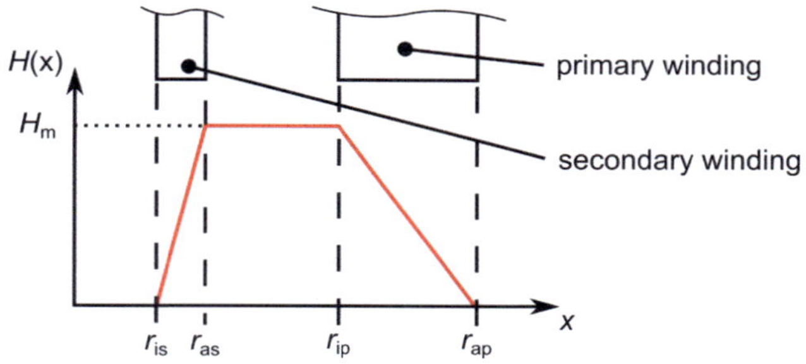

Figure 3.4: Magnetic field distribution in radial direction of the AC-SFCL, which is similar to the distribution of the magnetic field in transformers [Lei06]

The assumption of the distribution of the magnetic field across the windings is the same for conventional transformers [Lei06]. The corresponding function reads as follows

$$H(x) = \left\{ \begin{matrix} \frac{H_m}{r_{as} - r_{is}} x & for & r_{is} \leq x < r_{as} \\ H_m & for & r_{as} \leq x < r_{ip} \\ \frac{H_m}{r_{ap} - r_{ip}} (r_{ap} - x) & for & r_{ip} \leq x < r_{ap} \end{matrix} \right\} \tag{3.25}$$

In eq. (3.25) the magnetic field H_m is calculated using elliptic integrals as described above. This distribution of the magnetic field is used to calculate the stray inductance of the AC-SFCL.

The general correlation between the energy E_m stored in a magnetic field, the inductance L and the current I is

$$E_m = \frac{1}{2} \cdot L \cdot I^2 \tag{3.26}$$

The stored energy E_m can be as well calculated by integration over the magnetic field:

$$E_{\mathrm{m}} = \frac{\mu}{2} \cdot \int_{V} \mathrm{H}^2 \, \mathrm{dV} \tag{3.27}$$

Assuming, that the magnetic field is constant in axial and circumference direction and follows the radial distribution described in eq. (3.28) the energy stored in the magnetic stray field of across the windings can be expressed by

$$E_{\mathrm{m}} = \frac{\mu}{2} \cdot \mathrm{l}_{\mathrm{turn}} \cdot \mathrm{h}_{\mathrm{p}} \int_{0}^{\mathrm{b_w}} \mathrm{H(x)}^2 \, \mathrm{dx} \tag{3.28}$$

Herein l_{turn} is the average length of the turns, h the height of the winding and I the current conducted in the winding. If eq. (3.26) is inserted to eq. (3.28) the stray inductance results in

$$L_{\sigma} = \frac{\mu}{I_{\mathrm{p}}^2} \cdot l_{\mathrm{turn}} \cdot h_{\mathrm{p}} \int_{0}^{b_{\mathrm{w}}} H(x)^2 \cdot dx \cdot \tag{3.29}$$

Calculation of the inductance of a solenoid

In order to be able to calculate the inductance of solenoids with different diameters and heights an approach was developed which breaks down the calculation of the inductance of a solenoid winding to the calculation of the inductance of concentric, coupled current loops. Figure 3.5 shows two concentric current loops with different radii and axial distance.

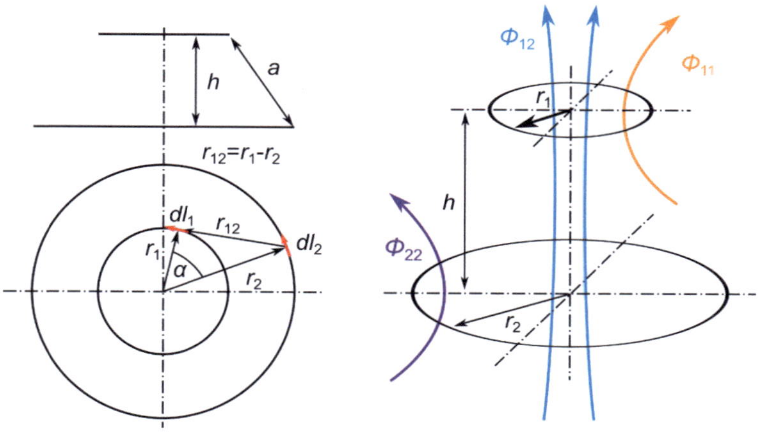

Figure 3.5: Two coaxial current loops with radii r_1 and r_2 and distance h

The approach to calculate the self-inductance of a current loop and the mutual inductance of two concentrically aligned current loops described in eq. (3.30) to eq. (3.47) is adopted from [Phi00]. Generally the magnetic flux Φ through a closed surface S, such as in a current loop, is the integration of the magnetic flux density B over the surface A [Fle10a]:

$$\Phi = \oint_S B \cdot dA \tag{3.30}$$

In eq. (3.30) the magnetic flux density B can be replaced by rot W, wherein W is the vector potential [Schw02]

$$\Phi = \oint_S rot\, W \cdot dA \tag{3.31}$$

By applying Stokes theorem [Fle10b], [Str06] eq. (3.31) can be written as

$$\Phi = \oint_{l_1} W \cdot dl_1 \tag{3.32}$$

With dl_1 being a line element in the current loop. In case of the magnetic flux generated by a current loop and linked with a second current loop eq. (3.32) becomes

$$\Phi_{12} = \oint_{l_2} W \cdot dl_2 \tag{3.33}$$

In the case of two coupled current loops the vector potential *W* can be expressed by [Phi00]

$$W = \frac{\mu \cdot I_1}{4\pi} \oint_{l_1} \frac{1}{a} \cdot dl_1 \tag{3.34}$$

Herein a is the distance between two current loops as shown in Figure 3.5. Inserting eq. (3.34) in eq. (3.33) leads to

$$\Phi_{12} = \frac{\mu \cdot I_1}{4\pi} \oint_{l_1} \oint_{l_2} \frac{dl_2 \times dl_1}{a} \tag{3.35}$$

The relation of magnetic flux Φ, inductance L and current I is in this case

$$\Phi_{12} = L_{12} \cdot I_1 \tag{3.36}$$

Inserting eq. (3.36) in eq. (3.35) gives an equation for the mutual inductance of both current loops

$$L_{12} = L_{21} = M = \frac{\mu}{4\pi} \oint_{l_1} \oint_{l_2} \frac{dl_1 \times dl_2}{a} \tag{3.37}$$

Herein dl_1 and dl_2 are two vectorial line elements. The scalar product of these line elements can be expressed as follows

$$dl_1 \times dl_2 = dl_1 \cdot dl_2 \cdot cos(\alpha) \tag{3.38}$$

Inserting eq (3.38) into eq. (3.37) gives

$$M = \frac{\mu}{4\pi} \oint_{l_1} dl_1 \oint_{l_2} \frac{cos(\alpha) \cdot dl_2}{a} \tag{3.39}$$

The integral over l_1 results in $2\pi r_1$. The integral over l_2 can be expanded to

$$dl_2 = r_2 \cdot d\alpha \tag{3.40}$$

Therefore the mutual inductance of two current loops is

$$M = \frac{\mu \cdot r_1 \cdot r_2}{2} \int_0^{2\pi} \frac{\cos\alpha \cdot d\alpha}{a} \tag{3.41}$$

Herein *a* is the distance of the two line elements of the current loops (compare Figure 3.5). The magnitude of this distance is

$$a = \sqrt{h^2 + r_{12}^2} \tag{3.42}$$

with

$$r_{12} = \sqrt{r_1^2 + r_2^2 - 2 \cdot r_1 \cdot r_2 \cdot cos\,\alpha} \tag{3.43}$$

Inserted in eq. (3.41) the mutual inductance can be expressed as

$$M = \frac{\mu \cdot r_1 \cdot r_2}{2} \int_0^{2\pi} \frac{cos\,\alpha \cdot d\alpha}{\sqrt{h^2 + r_1^2 + r_2^2 - 2 \cdot r_1 \cdot r_2 \cdot cos\,\alpha}} \tag{3.44}$$

If only one current loop is considered, that is to say the calculation of the self-inductance of one current loop, then

$$r_1 = r_2 \tag{3.45}$$

and

$$h = 0 \tag{3.46}$$

With eq. (3.45)and eq. (3.46) eq. (3.44) is simplified to

$$L = \frac{\mu \cdot r_1}{2} \int_0^{2\pi} \frac{cos\,\alpha}{\sqrt{2 - 2 \cdot cos\,\alpha}} \cdot d\alpha \tag{3.47}$$

Equation (3.47) allows the calculation of the self-inductance of one current loop. The cos-function in eq. (3.47) possesses two singularities, one at 0 and one at 2π, which must be paid attention during numerical calculation of the integral.

The winding of a solenoid coil can be approximated by a composition of single current loops as shown in Figure 3.6.

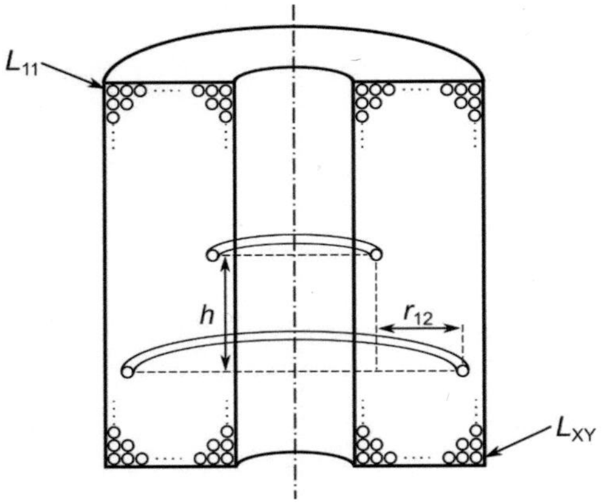

Figure 3.6: Cross-section of a solenoid winding divided into a finite number of current loops with two current loops indicated as in Figure 3.5

The self-inductance of each current loop can be calculated using eq. (3.47) and the mutual inductance between each current loop using eq. (3.44). The calculated values can be stored in a matrix with the inductance of the currents loops in the diagonal and the mutual inductance of each turn with the others at the respective position:

$$L_M = \begin{bmatrix} L_{11} & \cdots & L_{1Y} \\ \vdots & \ddots & \vdots \\ L_{X1} & \cdots & L_{XY} \end{bmatrix} \tag{3.48}$$

Summarization of the elements of the matrix L_M and dividing the result by the square number of current loops results in the inductance of the solenoid winding with one turn

$$L_{sol} = \frac{1}{(x \cdot y)^2} \sum_{n=1}^{x} \sum_{m=1}^{y} L_{nm} \tag{3.49}$$

The approach of dividing the cross-section of a solenoid winding into a finite number of current loops was defined as a function, which was eventually implemented as computational code. The implemented function is described in appendix A. The practical advantage of this approach is to have one callable function for any arbitrary solenoid geometry. With eq. (3.40) the inductance of the primary winding can be calculated by

$$L_p = L_{sol}\left(r_{ip}, r_{ap}, h_p\right) \cdot N_p^2 \tag{3.50}$$

Accordingly the inductance of the secondary winding can be calculated by

$$L_s = L_{sol}(r_{is}, r_{as}, h_s) \cdot N_s^2 \tag{3.51}$$

To fully describe the electrical behavior of the AC-SFCL it is necessary to calculate the stray inductances of both windings (compare Figure 3.3). As shown in Figure 3.7 (A) only the magnetic stray field, and therefore a stray flux, remains during normal operation of the AC-SFCL, because the coupled magnetic fields of both windings compensate each other. During fault operation the secondary winding is in normal conducting state as well, resulting in a magnetic field, and therefore in a linked magnetic flux, inside the secondary winding as shown in Figure 3.7 (B). To describe the electric behavior of the AC-SFCL it is necessary to translate this magnetic field distribution into corresponding elements of the equivalent circuit diagram.

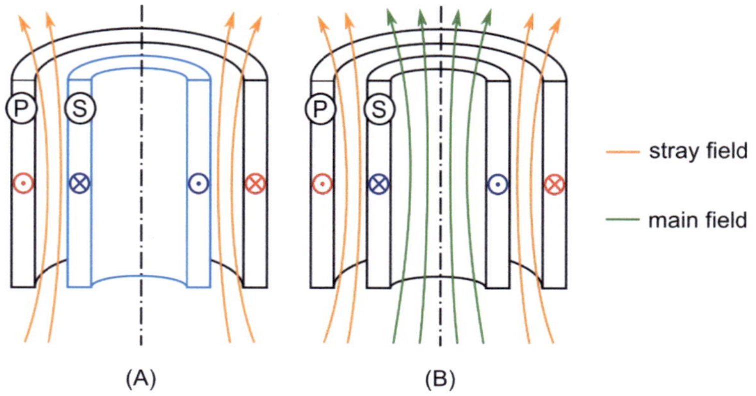

Figure 3.7: Magnetic field of the primary (P) and secondary (S) winding for (A) normal operation of the AC-SFCL (secondary winding in superconducting state) and (B) both windings are in normal conducting state.

The total magnetic flux Φ of each winding can be separated into a linked flux, which is coupled with the other winding and the stray flux, which is only associated with the respective winding:

$$\Phi_p = \Phi_{ps} + \Phi_{\sigma p} \tag{3.52}$$

$$\Phi_s = \Phi_{sp} + \Phi_{\sigma s} \tag{3.53}$$

Herein Φ_{ps} is the flux in the primary winding generated by the current in the secondary winding and Φ_{sp} the flux in the secondary winding generated by the current in the primary winding. Expanding eq. (3.52) and eq. (3.53) by the ratio of the respective number of turns *N* and the current *I* gives

$$\frac{\Phi_p \cdot N_p}{I_p} = \frac{\Phi_{ps} \cdot N_p}{I_p} + \frac{\Phi_{\sigma p} \cdot N_p}{I_p} \quad (3.54)$$

$$\frac{\Phi_s \cdot N_s}{I_s} = \frac{\Phi_{sp} \cdot N_s}{I_s} + \frac{\Phi_{\sigma s} \cdot N_s}{I_s} \quad (3.55)$$

The self-inductance of the solenoid windings can be expressed as the ratio of number of turns *N* times flux *Φ* and current *I* [Str06]. For the primary and secondary winding the corresponding equations are

$$L_p = \frac{N_p \cdot \Phi_p}{I_p} \quad (3.56)$$

$$L_s = \frac{N_s \cdot \Phi_s}{I_s} \quad (3.57)$$

For the mutual fluxes similar equations the equations read as follow:

$$M_{ps} = \frac{N_p \cdot \Phi_{ps}}{I_s} \quad (3.58)$$

$$M_{sp} = \frac{N_s \cdot \Phi_{sp}}{I_p} \quad (3.59)$$

Herein M_{ps} and M_{sp} are the mutual inductances between the windings, which is the same for both windings [Gri14]:

$$M = M_{ps} = M_{sp} \quad (3.60)$$

and hence

$$M = \frac{\Phi_{ps} \cdot N_p}{I_s} = \frac{\Phi_{sp} \cdot N_s}{I_p} \quad (3.61)$$

Similar to eq. (3.56) and eq. (3.57) the magnetic fluxes Φ_{ps} and Φ_{sp} determine the main inductances of the respective windings

$$L_{mp} = \frac{N_p \cdot \Phi_{ps}}{I_p} \tag{3.62}$$

$$L_{ms} = \frac{N_s \cdot \Phi_{sp}}{I_s} \tag{3.63}$$

The stray flux of each winding includes the magnetic field, which is not linked with the other winding. The corresponding equations are

$$L_{\sigma p} = \frac{N_p \cdot \Phi_{\sigma p}}{I_p} \tag{3.64}$$

$$L_{\sigma s} = \frac{N_s \cdot \Phi_{\sigma s}}{I_s} \tag{3.65}$$

Insertion of eq. (3.56), eq. (3.58) and eq. (3.64) in eq. (3.54) for the primary winding and eq. (3.57), eq. (3.59) and eq. (3.64) in eq. (3.55) gives

$$L_p = L_{mp} + L_{\sigma p} \tag{3.66}$$

$$L_s = L_{ms} + L_{\sigma s} \tag{3.67}$$

In eq. (3.66) and eq. (3.67) the total inductances L_p and L_s of the windings can be calculated using eq. (3.50) and eq. (3.51). Inserting in eq. (3.62) and eq. (3.63) in eq. (3.61) the mutual inductance M can be expressed as

$$M = L_{mp} \frac{N_s}{N_p} = L_{ms} \frac{N_p}{N_s} \tag{3.68}$$

The main inductances in eq. (3.68) can be replaced by eq. (3.66) and eq. (3.67)

$$\left(L_p - L_{\sigma p}\right) \cdot \frac{N_s}{N_p} = \left(L_s - L_{\sigma s}\right) \cdot \frac{N_p}{N_s} \tag{3.69}$$

Eq. (3.69) can be solved for the stray inductance $L_{\sigma p}$

$$L_{\sigma p} = L_p - \left(L_s - L_{\sigma s}\right) \cdot \left(\frac{N_p}{N_s}\right)^2 \tag{3.70}$$

For the composition of two concentrically aligned solenoids the total stray inductance is determined by

$$L_\sigma = L_{\sigma p} + L_{\sigma s} \cdot \left(\frac{N_p}{N_s}\right)^2 \tag{3.71}$$

In eq. (3.71) the total stray inductance can be calculated using eq. (3.29). Inserting eq. (3.70) into eq. (3.71) and solve for the inductance $L_{\sigma s}$ leads to

$$L_{\sigma s} = \frac{1}{2}\left[L_s + \left(\frac{N_s}{N_p}\right)^2 (L_\sigma - L_p)\right] \tag{3.72}$$

Similarly this can be done for the stray inductance $L_{\sigma p}$

$$L_{\sigma p} = \frac{1}{2}\left[L_p + L_\sigma - \left(\frac{N_p}{N_s}\right)^2 L_s\right] \tag{3.73}$$

With eq. (3.72) and eq. (3.73) the stray inductances $L_{\sigma p}$ and $L_{\sigma s}$ in the equivalent circuit diagram can be calculated, since the inductance of the primary winding L_p is given by eq. (3.50), the inductance of the secondary winding L_s by eq. (3.51) and the total stray inductance L_σ by eq. (3.29). The main inductances can be calculated using eq. (3.66) and eq. (3.67)

$$L_{mp} = L_p - L_{\sigma p} \tag{3.74}$$

$$L_{ms} = L_s - L_{\sigma s} \tag{3.75}$$

Calculation of resistances

The resistance of the primary winding can be calculated by

$$R_\mathrm{p} = N_\mathrm{p} \cdot \rho_\mathrm{p} \cdot \frac{A_\mathrm{cs}}{\pi \cdot (r_\mathrm{ip} + r_\mathrm{ap})} \tag{3.76}$$

Herein is ρ_p the conductivity and A_{cs} the cross-section of the conductor. The addition of the inner and outer radius of the primary winding (r_{ip} and r_{ap}) and multiplication with π is assumed as equivalent length of one turn. It is assumed, that the wire in the primary winding is made of many strands in order to minimize the necessary forces applied during the winding process. Additionally, this stranding makes the influence of the skin effect during operation at a nominal frequency of f_n = 50 Hz negligible.

The calculation of the resistance of the secondary winding R_s is similar to eq. (3.76)

$$R_s\ (T = T_c) = \rho_{nc} \cdot \frac{\pi \cdot (r_{is} + r_{as})}{A_{tape}} \cdot \frac{N_s}{n_{tot}} \tag{3.77}$$

Herein ρ_{nc} is the conductivity of the superconductor in normal conducting state (at $T = T_c$ for conservative estimation) and A_{tape} the cross-section of the *RE*BCO tape, n_{tot} is the total number of parallel *RE*BCO tapes in secondary winding for $N_s = 1$. In case of more than one turn ($N_S > 1$) the actual number of tapes connected in parallel n_{par} is:

$$n_{par} = \frac{n_{tot}}{N_s} \tag{3.78}$$

In order to calculate the impedance of the AC-SFCL it is necessary to transform the resistance R_{stab} to the primary side

$$R'_s = c_f \cdot R_s \tag{3.79}$$

If eq. (3.4), eq (3.77) and eq (3.78) are inserted in eq (3.79) the resistance of the secondary winding transformed to the primary side is

$$R'_s = \rho_{nc} \cdot \frac{\pi \cdot (r_{is} + r_{as})}{A_{tape}} \cdot \frac{N_p^2}{n_{par}} \tag{3.80}$$

This means, that the resistance R'_s is independent of the number of turns of the secondary winding N_s or the circuitry of the *RE*BCO tapes. As a result the wire configuration of the secondary winding can be freely chosen. As described in chapter 2.4 one option is to short-circuit each *RE*BCO ring with itself, resulting in a secondary winding with one turn ($N_s = 1$). Another option would be to build modules with several turns and stack the modules to shield the primary winding. Either way the resistance R'_s of the secondary winding will be the same.

3.3 Calculation of AC losses

Losses in superconductors caused by alternating currents can be categorized in two separate ways. On the one side there are hysteresis losses, which are caused by flux flow in the superconductor and eddy current losses in the normal conducting layers of the superconductor.

These loss mechanisms are determined by the material properties and their geometry. In superconductors the magnetic field is penetrating the superconductor in quantized flux lines. If the superconductor is exposed to an alternating magnetic field these flux lines are moving. This movement requires energy and is the reason for these losses. To calculate these AC-losses the critical current, the magnetic field and the geometry must be known.

The eddy current losses occur in the normal conducting layers of the tapes. Exposed to an alternating magnetic field a voltage will be induced in these normal conducting layers and cause a current flow. For the calculation of the eddy current losses the geometry (cross-section), the resistivity or conductivity respectively and the magnetic field must be known.

Another categorization of AC-losses is to separate them by the origin of their magnetic field. The magnetic field can be caused by the alternating transport current in the superconductor itself or alternating currents conducted in nearby conductors.

Due to the anisotropic behavior of the superconducting material, the direction of the magnetic field (parallel or perpendicular) and the geometry of the superconductor are parameters for the calculation of these AC losses.

Calculation of eddy current losses

Eddy current losses can be calculated by the following equation [NB88]:

$$P_e = \frac{\pi^2 \cdot (f \cdot B_\perp)^2 \cdot b_{sc}^3 \cdot h_{stab} \cdot l_{sc}}{6 \cdot \rho_{nc}} \tag{3.81}$$

Herein f_n is the nominal frequency, $B_\perp$ the perpendicular magnetic field penetrating the *RE*BCO tape, l_{sc} the length of the *RE*BCO tape and ρ_{nc} the specific resistance of the normal conducting layer. According to [NAJ04] eddy currents caused by a perpendicular magnetic field do have a relevant impact in low fields on AC-losses. Due to the width of the *RE*BCO tape the parallel magnetic field component is negligible. Eddy current losses caused by the self-field are as well negligible [ONO04].

Calculation of hysteresis losses caused by self-field

According to [Nor69] the hysteresis losses of a *RE*BCO conductor with rectangular cross-section can be calculated using the following equation:

$$P_{hs} = \frac{I_c^2 \cdot l_{sc} \cdot \mu_0}{\pi}\{(1-F) \cdot ln(1-F) + (1+F) \cdot ln(1+F) - F^2\} \tag{3.82}$$

wherein F is the ratio between the critical current I_c and the conducted peak current I_m

$$F = \frac{I_m}{I_c} \tag{3.83}$$

The critical current I_c must hereby incorporate the magnetic field at the position of the tape within the AC-SFCL.

Calculation of hysteresis losses caused by external magnetic-field

The hysteresis losses caused by an external, alternating, perpendicular magnetic field can be calculated using the following equation according to [Bra94] and [Bl93]:

$$P_{he\perp} = \pi \cdot f_n \cdot l_{sc} \cdot b_{sc}^2 \cdot \frac{1}{\mu_0} \cdot B_\perp \cdot g\left(\frac{B_\perp}{B_c}\right) \tag{3.84}$$

Wherein the function g is defined as follows

$$g(x) = \frac{1}{x} \cdot \left[\frac{2}{x} \cdot ln(cosh\, x) - tanh\, x\right] \tag{3.85}$$

The critical magnetic B_c field can be estimated by

$$B_c = \frac{\mu_0 \cdot I_c}{\pi \cdot b_{sc}} \tag{3.86}$$

The hysteresis loss caused by an external, alternating parallel magnetic field can be calculated using the following equations according to [MW01]:

$$P_{he\parallel} = \begin{cases} \frac{2 \cdot f_n \cdot l_{sc} \cdot A_{pen}}{3 \cdot \mu_0 \cdot B_c} B_\parallel^3, & B_\parallel \leq B_{pen} \\ \frac{2 \cdot f_n \cdot l_{sc} \cdot A_{pen} \cdot B_{pen}}{3 \cdot \mu_0} (3 \cdot B_\parallel - 2 \cdot B_c), & B_\parallel \geq B_{pen} \end{cases} \tag{3.87}$$

Herein A_{pen} is the surface penetrated and B_{pen} the full penetration field, which can be calculated by

$$B_p = \mu_0 \cdot I_c \tag{3.88}$$

The total AC losses can be calculated by summation of all components for each tape in the secondary, superconducting winding and summation of losses of all *RE*BCO tapes

$$P_{ac} = n_{par} \cdot P_{hs} + \sum_{n=1}^{n=n_{par}} (P_{he\perp} + P_{he\parallel} + P_e) \tag{3.89}$$

Herein n is the iteration variable over all *RE*BCO tapes in the secondary winding.

3.4 Design method for an Air Coil Superconducting Fault Current Limiter

The design method as shown in Figure 3.8 was set-up in such a way, that the calculation of the primary and the secondary winding can be performed independently. The advantage of this approach is, that it covers two possible applications.

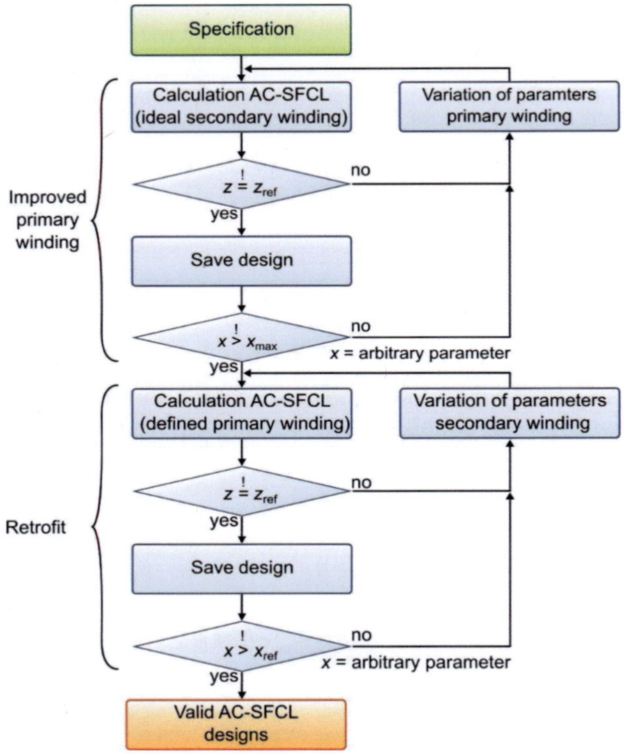

Figure 3.8: General approach of the proposed design method. The design process is separated into two major parts, the calculation of a design with an improved primary winding and the insert. This allows independent the calculation of the retrofit of an air core reactor and the design of an AC-SFCL with improved primary winding.

The first part is the design of an insert in order to retrofit an air core reactor. In this case the dimensions of the primary winding are already given by the air core reactor. By variation of the properties of the secondary winding, mainly the properties of the *RE*BCO tape, the impedance z_{lim} during fault and quench behavior can be influenced. The impedance during nominal operation z_n can be adjusted by variation of the winding distance d_w.

The second application is to build an AC-SFCL design with improved primary winding. Therefore the geometrical parameters of the primary winding are varied in order to adjust the electrical parameters, mainly the impedances, to specified values. In this case the parameters of the secondary winding are set constant.

3.4.1 Retrofitting an Air Core Reactor

In case of a retrofit the geometry of the primary winding is already defined by the dimensions of the air core reactor. The critical current I_c and the thickness h_{stab} of the stabilization of the *RE*BCO as well as the distance d_w between the windings are variables. The geometry and the variables, defined in a reasonable range, serve as specification for the design process shown in Figure 3.9. Before iterating over the variables in the defined range the magnetic field within the primary winding is calculated in step 2 using elliptic integrals as described in chapter 3.2. The magnetic field is calculated in axial and radial direction of the solenoid. This corresponds to the perpendicular and parallel direction of the magnetic field related to the *RE*BCO tapes. For the further calculation the maximum magnetic field component is considered, which gives a lower I_c for the *RE*BCO tapes as a worst case assumption.

In calculation step 3 the critical current I_c, thickness of stabilizer h_{stab} and winding distance d_w are varied. These variable parameters have a direct influence on the resistances and reactance of the equivalent circuit diagram and therefore on the impedance during normal operation and fault condition. The critical current I_c of the SC-tape determines the total amount of SC-tape needed in order to carry the induced current in the secondary winding safely. Since the thickness of the stabilizer of each SC-tape is assumed to be the same, the amount of parallel tapes affects the resistance during fault operation as well. A reasonable value range for critical current I_c is given in Table 3.3.

The stabilizer thickness h_{stab} is the main parameter determining the resistance of a *RE*BCO tape during fault operation. Usually it is a copper layer supported by a small layer of silver. A reasonable range of values for copper stabilization is given in Table 3.3.

The winding distance d_w determines the magnetic stray field. It is the major variable parameter influencing the impedance during normal operation. With increasing winding distance the magnetic stray field increases and therefore the impedance during normal operation. The objective is to design the distance as small as possible, but at the same time ensure thermal and electrical insulation. The maximum value should be around 10% of the inner diameter of the primary winding. During one calculation cycle all variable parameters remain constant.

In calculation step 4 the number of parallel *RE*BCO tapes n_{par} is calculated using eq. (3.90) and eq. (3.91).The number of parallel tapes n_{par} is estimated assuming ideal coupling between the windings

$$\frac{N_\mathrm{p}}{N_\mathrm{s}} = \frac{I_\mathrm{max}}{I_\mathrm{n,peak}} \cdot n_\mathrm{par} \tag{3.90}$$

wherein

$$I_\mathrm{max} = I_\mathrm{c}(B,T) \cdot k_\mathrm{I} \tag{3.91}$$

The total number of tapes n_tot defines the cross-section A_s of the second winding and is calculated in step 5 using

$$A_\mathrm{s} = A_\mathrm{tape} \cdot n_\mathrm{tot} \cdot f_\mathrm{fs} \tag{3.92}$$

$$b_\mathrm{s} = \frac{A_\mathrm{s}}{h_\mathrm{p}} \tag{3.93}$$

Herein h_p is the height of the primary winding, which needs to be shielded, f_fs the fill-factor of the secondary winding and b_s the thickness of the secondary winding and n_tot can be calculated using eq. (3.78). With the thickness b_s and the winding distance d_w the outer and inner radius of the secondary winding can be calculated

$$r_\mathrm{as} = r_\mathrm{ip} - d_\mathrm{w} \tag{3.94}$$

$$r_\mathrm{is} = r_\mathrm{as} - b_\mathrm{s} \tag{3.95}$$

When all geometic parameters of the secondary winding have been calculated both windings are properly defined. The inductance of the secondary winding L_s in step 6 can be calculated using eq. (3.51).

The engineering current density of each winding are calculated using

$$j_\mathrm{ep} = \frac{I_\mathrm{n}}{h_\mathrm{p} \cdot (r_\mathrm{ap} - r_\mathrm{ip})} \tag{3.96}$$

$$j_\mathrm{es} = \frac{I_\mathrm{c}(B,T) \cdot n_\mathrm{par}}{h_\mathrm{s} \cdot (r_\mathrm{as} - r_\mathrm{is})} \tag{3.97}$$

Assuming a homogeneous distribution of the engineering current densities over the cross-section of the windings they can be used to calculate the magnetic field in the stray gap using elliptical integrals as described in chapter 3.2 and superpose the components of each winding. It is assumed, that the magnetic field in circular/phi-direction is constant as well as in axial direction (step 7). The distribution in radial direction is shown in Figure 3.4. Integration of this magnetic field using eq. (3.29) gives the total stray inductance L_σ. Based on the magnetic field distribution and the corresponding total stray inductance L_σ the stray inductance and main inductance of each winding can be calculated using eq. (3.72) through eq. (3.75) (step 8).

To fully describe the equivalent circuit diagram the resistances must be determined in step 9. The resistance of the primary winding R_P is assumed to be given (it can be calculated using eq. (3.76) otherwise) and the resistance of the secondary winding R_S or R_S' respectively is calculated using eq. (3.77) and eq. (3.80). With all elements of the equivalent circuit diagram defined the impedance for normal operation and fault operation can be calculated and compared with the previously defined reference values.

If the calculated impedance z_n is equal or lower than a reference impedance z_{nref} the current design is saved, if not the calculation process is started again with a new set of variable parameters. After a feasible design matches the impedance criterion and is saved the variable parameters are varied consecutively until defined maximum values are reached. If all variable parameters are having passed their respective value range the calculation process ends.

If the calculated values for impedance meet the defined condition the calculated design is stored as feasible and the calculation can be started with a new set of variable parameters, if not the iteration is started with a new set of variable parameters immediately. This process continues until all variable parameters have been processed.

The result of this process are all designs, which match the specified impedance during normal operation. Within these results it is now possible to find designs, which are optimized with respect to certain parameters, such as losses, volume or *RE*BCO conductor demand for example. This can be achieved by searching for the design with the lowest losses, volume or *RE*BCO conductor demand.

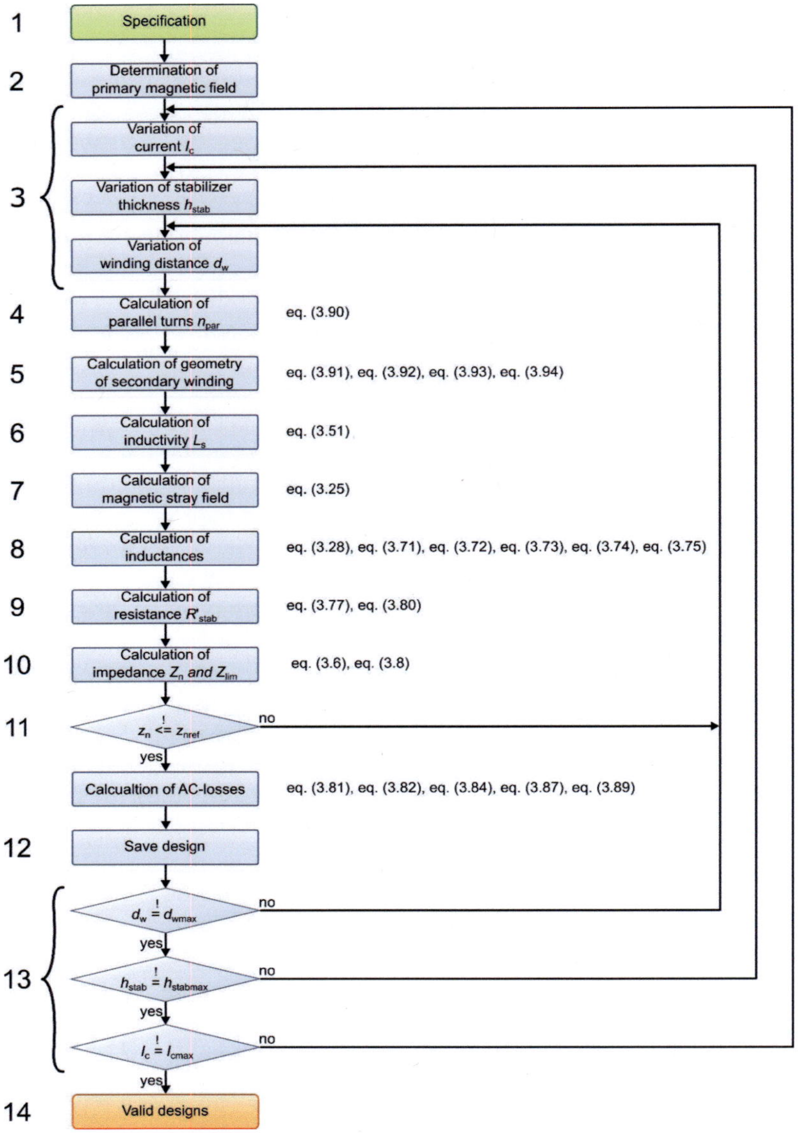

Figure 3.9: Flow diagram for calculation of the retrofit. Assuming constant parameters for the primary winding a secondary winding is calculated. If the impedance of the design during normal operation the AC-losses of the AC-SFCL design are calculated and the design is stored.

3.4.2 Optimized design of an Air Coil Superconducting Fault Current Limiter

The impedance of the AC-SFCL during normal operation is mainly determined by the stray field across the windings. For the retrofit the only possibility to minimize the stray field is to narrow the gap between the windings. However, another option is designing the primary winding from scratch. By variation of the geometry of the primary winding the magnetic field distribution and magnitude can be altered and therefore the magnetic stray field over the primary winding minimized. The basic specification remains the same compared to the retrofit: Apparent power S_n, voltage V_n and impedance z_{lim} must be defined. The approach in this case is to vary the geometrical parameters inner radius r_{ip} and height h_p and determine iteratively the appropriate number of turns N_p of the primary winding in order to match the defined reference impedance $z_{ref} = z_{lim}$ or the inductance L_{lim} (eq. (3.12)) respectively. Figure 3.11 shows the design method to calculate an optimized primary winding for an AC-SFCL.

After the specification in calculation step 1, the inner radius r_{ip} and the height h_p are varied in a given interval and increment (step 2).

The number of turns N_p is determined in an iterative way (step 2 to step 5), since for a given inductance L_{lim} the corresponding number of turns N_p needs to be found. The calculation is started with an initial number of turns $N_p = N_{init}$ and the necessary cross section is calculated taking into account the fill factor of the primary winding f_{fp}. This calculation follows

$$A_p = f_{fp} \cdot N_p \cdot A_c \tag{3.98}$$

wherein A_c is the cross-section of each single conductor or turn and A_p is the cross-section of the primary winding. The fill factor f_{fp} is defined as the ratio of conductor cross-section to the total cross-section of the primary winding. The outer radius of the primary winding can be calculated

$$r_{ap} = \frac{A_p}{h_p} - r_{ip} \tag{3.99}$$

With eq. (3.99) all geometrical parameters are defined and the inductance of the primary winding L_p can be calculated. If the calculated inductance L_p is higher than the inductance L_{lim} the calculation proceeds with the next step. If this is not the case the number of turns N_p is stored as N_{min} and a new number of turns N_{max} is calculated by increasing N_{min} (e.g. doubling N_{min}). The above described calculation of cross-section A_p, radius r_{ap} and inductance L_p is repeated until L_p is larger than L_{lim} (step 3).

If this is the case then there is a number of turns N_{min}, which leads to a lower inductance than L_{lim} and a number of turns N_{max}, which leads to a higher inductance than

L_{lim}. The number of turns N_p, which corresponds to $L_p = L_{lim}$, must be found between N_{min} and N_{max}.

In step 4 the average of N_{min} and N_{max} is calculated and verified, if this leads to an inductance L_p, which converges to L_{lim}.

If this is not the case, this process is repeated (step 5). Hereby either N_{max} is reduced or N_{min} is increased until either one of them leads to an inductance L_p, which converges with L_{lim}. This approach is qualitatively shown in Figure 3.10. The advantage of this approach is, that the inductance L_p is calculated within a few steps, which is practically faster than looping through the number of turns.

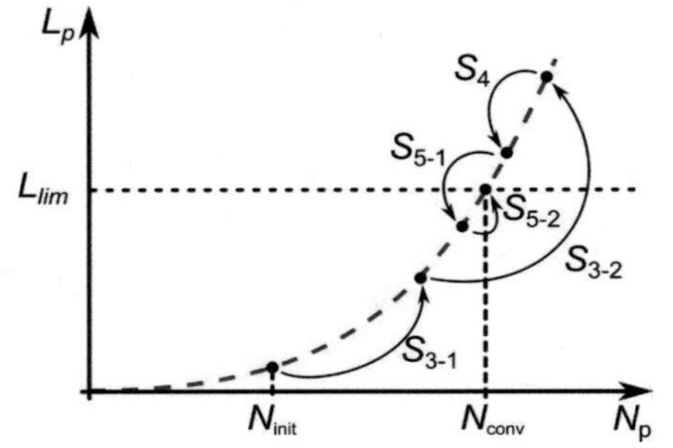

Figure 3.10: Iterative approach to calculate the inductance L_p of the primary winding The number of turns is varied in discrete steps until the calculated impedance L_p matches the impedance necessary for fault current limitation

To fully characterize the primary winding the resistance is calculated in step 6 with the specified conductor cross-section and its electrical resistivity at the foreseen temperature (usually RT) using eq. (3.62).

In step 7 the impedance of the primary winding can then be calculated by

$$Z_\mathrm{p} = \sqrt{R_\mathrm{p}^2 + \left(2\pi \cdot f_\mathrm{n} \cdot L_\mathrm{p}\right)^2} \tag{3.100}$$

and the reference impedance z_{lim} according to eq. (3.10). The calculated impedance Z_p is compared to the specified impedance Z_{lim}. If both impedances match, the calculation is continued assuming an ideal secondary winding. This means, that the secondary winding is assumed to ideally shield the primary wining (fill factor $f_{fs} = 1$), has a high resistance (stabilizer thickness of the *RE*BCO tapes $h_{stab} = 2$ µm) and the critical current I_c of the *RE*BCO tapes is sufficiently high to carry the induced current using only one layer of *RE*BCO tapes in step 9.

At last, the impedance during normal operation is calculated assuming, that the resistance R'_{sc} is negligible and only the main inductance L_m and the stray inductance $L'_{\sigma s}$ are in parallel (step 10). The design is considered as feasible, saved and the calculation starts with a new set of parameters until variable parameters inner radius r_{ip} and height h_p reach their defined maximum value.

Finally, there is a set of geometrical parameters, which can be found in the searched range of the geometrical parameters and meet the specifications. Within these results an optimum can be found by the maximum ratio of impedance during fault to impedance during normal operation

$$f_z = \frac{Z_p}{Z_n} \tag{3.101}$$

With a given ration f_z, it is now possible to search for the most compact geometry or smallest volume in the calculated results. If the overall dimensions are a constraint, for example if a maximum height or maximum diameter should not be exceeded, the design with the maximum f_z can be found as well existing in the calculated results.

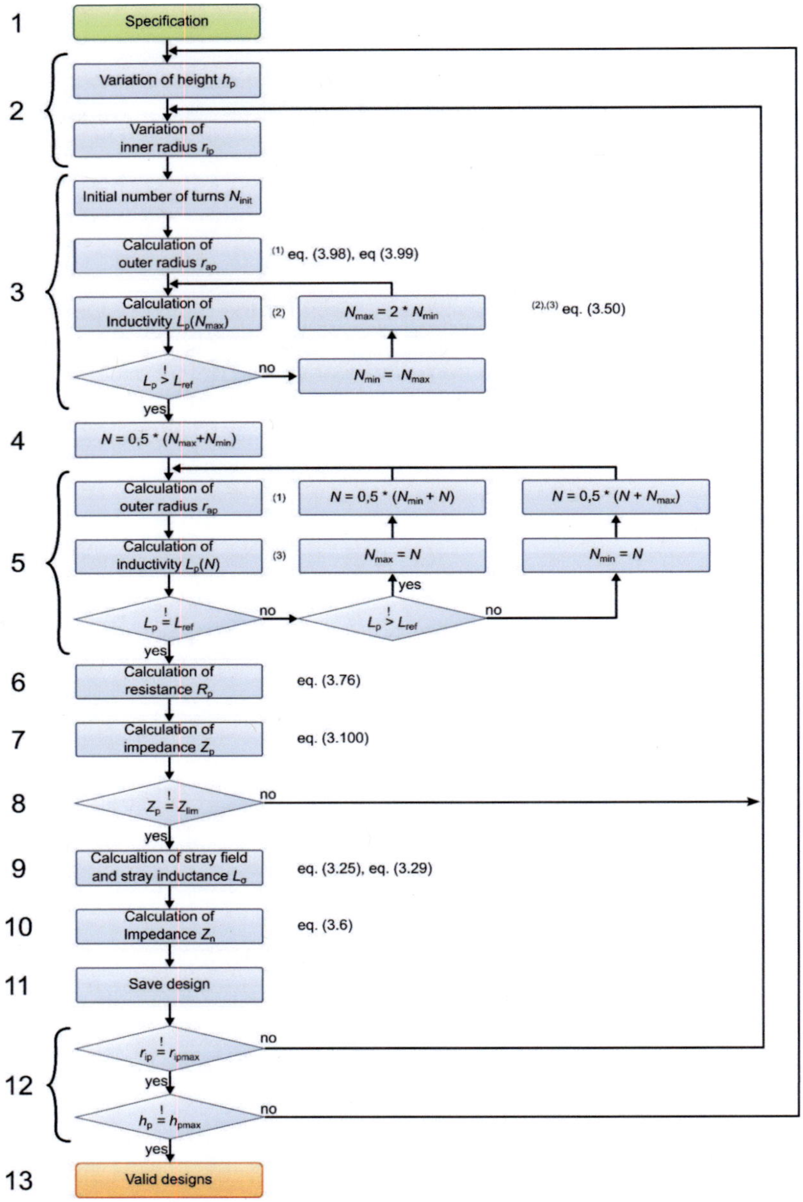

Figure 3.11: Flow diagram for optimized design. With a set of geometrical parameters the number of turns N are calculated to match the reference impedance L_{ref} and the impedance Z_{lim} assuming constant parameters for secondary winding.

3.5 Summary

This chapter described the design and the basic functionality of an AC-SFCL. Equations have been introduced in order to calculate the major electrical parameters of the equivalent circuit diagram depending on the geometry and the properties of the *RE*BCO tape. Therefore a method for the calculation of the magnetic stray field, which determines the stray inductances and hence the impedance during normal operation, with elliptic integrals has been adopted. A method for the calculation of the inductance of a solenoid has been introduced. This method allows the calculation of the inductance of solenoids with arbitrary cross-section and can therefore be used to calculate the inductance of the primary and the secondary winding.

The equations have been compiled into a design method. This allows a computational approach to design an AC-SFCL. Hereby the design method was divided into two separate parts. This allows to either retrofitting an air core reactor with a secondary winding or designing the AC-SFCL from scratch with an improved primary winding.

For the retrofit the variation of the properties of the *RE*BCO tape, critical current I_C, stabilizer thickness h_{stab} and distance of windings d_W allows to tune the electrical parameters and therefore the impedance z_n during normal operation and the impedance z_{lim} during current limitation.

Designing an AC-SFCL with improved primary winding allows improving the electrical parameters by variation of the geometrical parameters inner diameter d_{ip} and height h_p of the primary winding. In this case the geometry can be optimized in such a way, that the stray field between the windings is minimized and hence the impedance during normal operation.

Both design methods require the specification of the electrical parameters and specification of the superconducting *RE*BCO tape used in the secondary winding.

The design method was successfully used for the design of an AC-SFCL demonstrator described in chapter 4 and for the designs for medium and high voltage AC-SFCLs presented in chapter 6.

4 Design, manufacturing and test of a 60 kVA, 400 V, z = 6% demonstrator

This chapter describes the design, build and test of a 60 kVA, 400 V, z = 6% AC-SFCL demonstrator. As a prerequisite measurements are described, which were performed to characterize *RE*BCO tapes and to investigate their applicability for use in the demonstrator. Superconducting rings were manufactured from pieces of the *RE*BCO tapes and placed in a small coil for quench investigation. This small coil was also used to investigate the manufacturing process of the demonstrator and to address design challenges at an early stage. Additionally, the experimental setup was tested and adapted to the subsequent tests with the demonstrator.

The experimental results served as input parameters for the design of the demonstrator together with the equations and methods described in chapter 3. The most compact design was chosen, which meets the electrical specification. The demonstrator was investigated in three operational modes: Normal operation, fault current limitation and recovery under load. The first two operational modes proved the principle of the AC-SFCL concept as well as allowed verification of the design equation and methods.

4.1 Preliminary investigations of single, short-circuited superconducting tapes

Before the *RE*BCO tapes were used in the manufacturing process of the secondary winding of the demonstrator several measurements were performed to characterize them:

- Measurement of the resistance depending on temperature $R(T)$ of small samples allows determination of the critical temperature T_c and the progression of the resistance with increasing temperature of the *RE*BCO tape in normal conducting state.
- The critical current $I_c(B,T,\alpha)$ was measured on small samples for all angles and magnetic fields up to 600 mT. This allowed a prediction of the current carrying capacity of the tape in the demonstrator.
- The critical current I_c was measured at self-field on *RE*BCO tapes with 1.47 m piece length to prequalify the tapes for the demonstrator with respect to $_c$ homogeneity

In advance of manufacturing and testing of the AC-SFCL demonstrator a test coil was built to quench single superconducting rings made of *RE*BCO tape. The rings were fabricated by soldering the *RE*BCO tape at the ends using a short piece as interconnector. To manufacture the soldered joint an apparatus was built and the resistances of the soldered connections of small samples were measured. The quench-experiments with the test coil and single superconducting *RE*BCO rings allowed the adaption of the measurement equipment, such as a reliable connection of voltage taps on the *RE*BCO rings and protection circuits for sensitive measurement equipment.

4.1.1 Characterization of Superconducting Tapes

All measurements described in this chapter used two different *RE*BCO tapes. The data provided by the manufacturer is summarized in Table 4.1.

Table 4.1: Data of the *RE*BCO tapes

Parameter	Tape A	Tape B
Manufacturer Nr.	M3-876-6 SCS12050-AP	M3-913-1 SCS12050-AP
Tape width	12 mm	12 mm
Total thickness	94 µm	155 µm
Thickness REBCO layer	1 µm	1 µm
Mechanical stabilization: Hastelloy	50 µm	50 µm
Electrical stabilization: Silver	2 µm	2 µm
Electrical stabilization: Copper	40 µm	100 µm

Measurement of resistance depending of temperature $R(T)$

Measuring the resistance as a function of the temperature is important for the limiting behavior and for determination of the critical temperature T_c.

The measurement principle is shown in Figure 4.1. A sample of a superconducting tape is cooled down to 77 K in a LN_2 open bath cryostat and a current source is used to conduct a DC current of I_{meas} = 20 mA in the sample. As soon as the sample is in superconducting state the voltage drop V_{meas} over the sample is not measurable any more. A second current source was used to conduct a heating current I_{heat} in a resistance. The joule heat in the resistor warms up the sample at a given rate. The currents I_{heat} and I_{meas} are set using a Labview program running on a conventional PC, as it is the case for the voltage V_{meas} and a temperature signal.

Figure 4.2 shows the results for the *RE*BCO tapes A and B. Tape A shows a critical temperature of T_c =91 K and a resistance of 6.9 mΩ/m after transition to normal state. Tape B shows a critical temperature of 90.3 K and a resistance of 3 mΩ/m after transition to normal state. The reason for the higher resistance of tape A is the thinner copper layer compared to tape B as well as for the stronger linear gradient of resistance with increasing temperature.

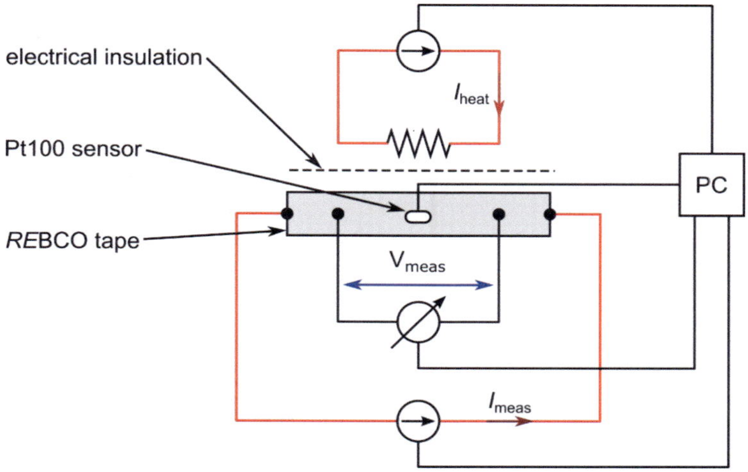

Figure 4.1: Diagram of the setup to measure the resistance depending on temperature $R(T)$. The *RE*BCO tape is cooled down to 77 K and the slowly heated up. During the heating process a small current I_{meas} and the voltage drop across the tape V_{meas} are used to calculate the resistance $R(T)$.

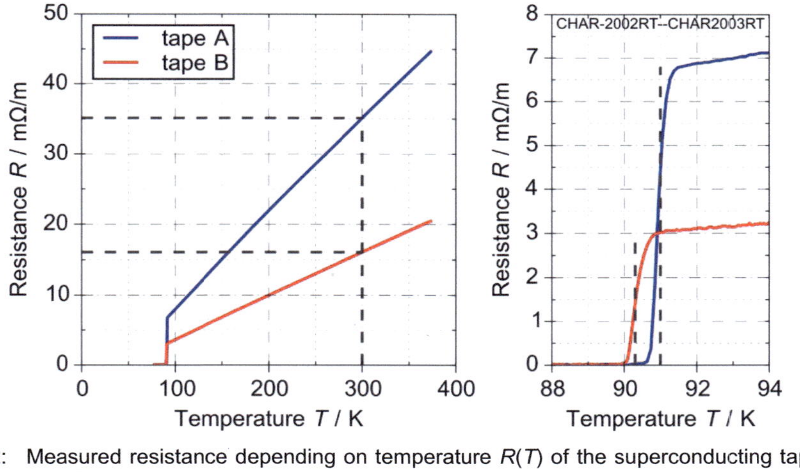

Figure 4.2: Measured resistance depending on temperature $R(T)$ of the superconducting tape A and tape B in the measured range

At room temperature (300 K) tape A reaches the calculated resistance assuming a resistivity of ρ = 0.0172 Ωm/mm². Tape B shows a higher resistance than calculated, which may be explained by geometry deviations of the tape layers, but was still within 11%. The results of the $R(T)$ measurements are summarized in Table 4.2.

Table 4.2: Summary of results of $R(T)$-measurement

Parameter	Tape A	Tape B
Critical temperature T_c	91 K	90.3 K
Minimum resistance in normal state	6.9 mΩ/m	3 mΩ/m
Measured resistance at 300 K	35.12 mΩ/m	16.08 mΩ/m
Theoretical resistance at 300 K	35.83 mΩ/m	14.33 mΩ/m

The resistance after transition to normal state was used to calculate the impedance during normal operation of the AC-SFCL demonstrator.

Measurement of critical current depending on magnetic field and angle Ic(B,T)

Each *RE*BCO tape in the AC-SFCL is subject to a different magnetic field configuration. While at the center of the solenoid the magnetic field is almost exclusively in parallel to the SC-tape the perpendicular field component increases towards the ends of the coil.

*RE*BCO tapes show an anisotropic behavior of the critical current I_C depending on the magnetic field.

Figure 4.3 shows the setup for measuring the critical current $I_c(B,T)$. The magnetic field is generated by two Helmholtz coils. This arrangement is suitable to generate a homogeneous magnetic field.

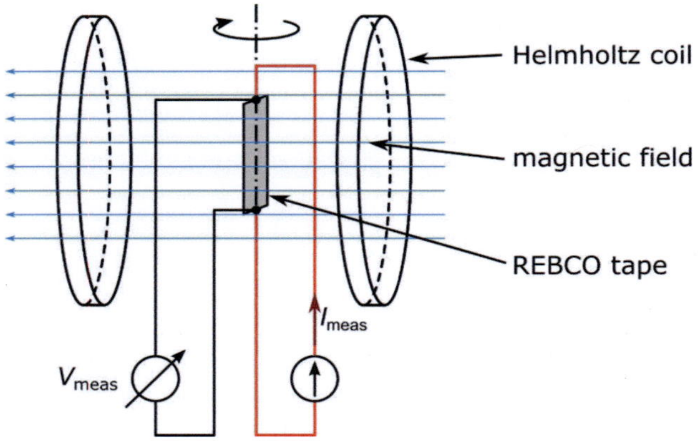

Figure 4.3: Schematic of the setup for measuring the critical current $I_c(B,T)$. A homogeneous magnetic field is generated using two Helmholtz coils. he *RE*BCO tape is rotated within the magnetic field.

A sample of 5 cm piece length was cooled down to 77 K in an open bath cryostat and placed between the coils. The magnetic field was ramped up and a current applied. The current was increased until the sample quenches. After the quench the current and the magnetic field were then shut down. The sample was rotated by an angle of 15°. This procedure was repeated until the sample was rotated by an angle of 360° in total.

In Figure 4.4 and Figure 4.5 the dependency of critical current I_c on the magnetic flux density B is shown for tape A and tape B.

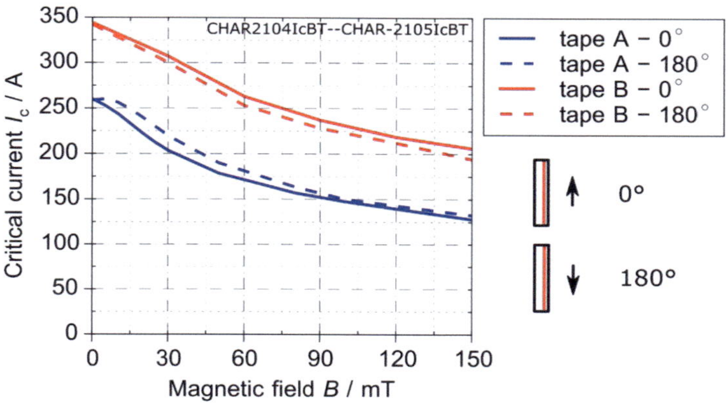

Figure 4.4: Measured critical current I_c depending on magnetic field density for parallel field

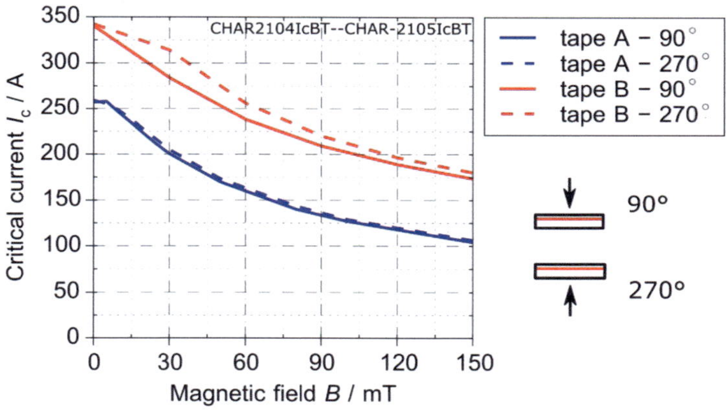

Figure 4.5: Measured critical current I_c depending on magnetic field density for perpendicular field

The magnetic field in Figure 4.4 was applied parallel to the tapes and the tapes were positioned at 0° and at 180°. The magnetic field in Figure 4.5 was applied perpendicular to the tapes and the tapes were positioned at 90° and at 270°.

Tape A and tape B show the expected decline of the critical current I_c with increasing magnetic field density B. Depending on the orientation of the tapes (0° or 180°) the critical current I_c deviates 11.4% at maximum for tape A and 6% at maximum for tape B.

In case of perpendicular field the critical current I_c decreases as well with increasing magnetic field density B.

The results for critical current I_c depending on the angle are shown in Figure 4.6 for tape A and tape B, both at a magnetic field density of B = 30 mT. The critical current I_c of tape A varies between 188 A at 50° and 231 A at 150°. For tape B the critical current I_c varies between 284 A at 240° and 346 A at 30°.

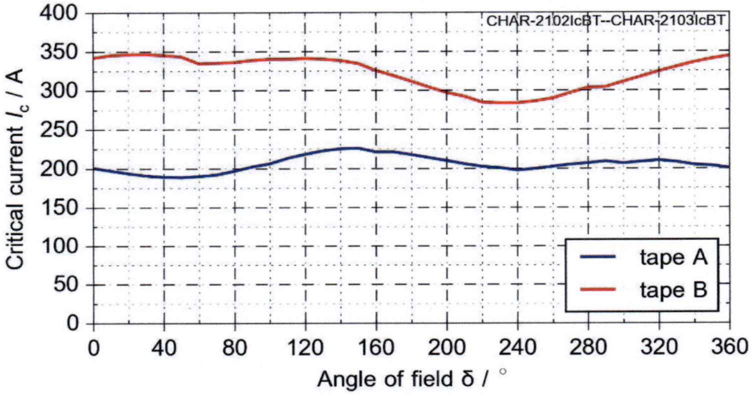

Figure 4.6: Measured critical current I_c depending on the angle of the magnetic field for tape A and tape B for a magnetic field density of B = 30 mT

Both tapes do not follow a particular shape in terms of dependency of critical current I_c of magnetic field angle. The reason for this behavior is the doping of the superconductor material during the manufacturing process in order to pin of the flux vortices and prevent their movement within the magnetic field.

Measurement of critical current Ic of superconducting tapes in self-field

The critical current I_c of *RE*BCO tapes is not homogeneous distributed and varies along the length of the tape. Reasons for this are inhomogeneity, such as local misalignment of the atomic lattice and variation of the thickness of the superconducting layer.

It was therefore necessary to investigate the distribution of the critical current I_c of each *RE*BCO tape in order to determine their applicability in the demonstrator. For every *RE*BCO tape the critical current I_c and the homogeneity of the critical current I_c along the tape was measured. Therefore the *RE*BCO tape was divided into 10 sections of 15 cm length each. The voltage drop of each interval was measured as well as the voltage over the whole tape.

The setup for the I_c-measurement is shown in Figure 4.7. A current source was used to control a DC current I_{meas} in the *RE*BCO tape. The voltage drop was measured at the terminals and across the whole *RE*BCO tape. Additional voltage taps were attached to the *RE*BCO tape along the length in 10 sections.

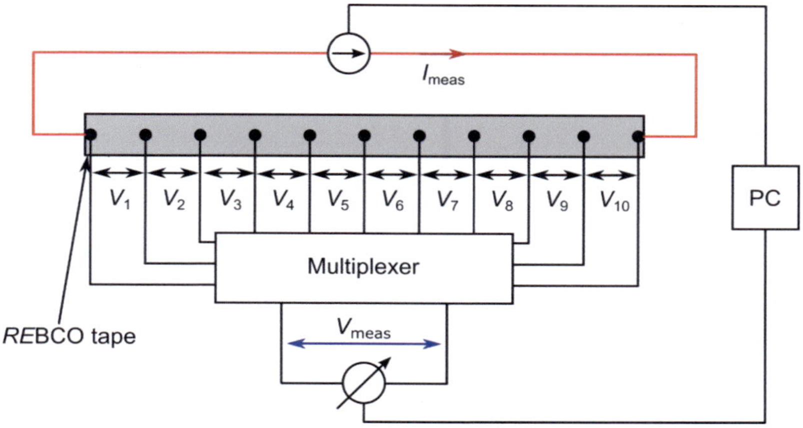

Figure 4.7: Setup to measure critical current distribution of each *RE*BCO tape intended for the use in the AC-SFCL demonstrator. The current is increased in discrete steps and the voltage drop of each section is measured after each current increase.

After an offset measurement to minimize measurement errors due to temperature the current was carefully ramped up in discrete steps. After each current step the voltage was measured. Hereby a multiplexer was used, to automatically measure all 10 voltage sections. Each voltage drop V_{meas} itself was measured using a Nanovoltmeter. The current I_{meas} was set via PC using a Labview program. The measured voltage V_{meas} is recorded by the same software.

Figure 4.8 shows the fixture for the *RE*BCO tapes together with the voltage contacts and current leads. The *RE*BCO tape was bended circularly with a diameter of 470 mm, which is the diameter of the secondary winding of the demonstrator (compare chapter 4.3). The setup shown in Figure 4.8 was emerged completely in LN_2 under boiling conditions.

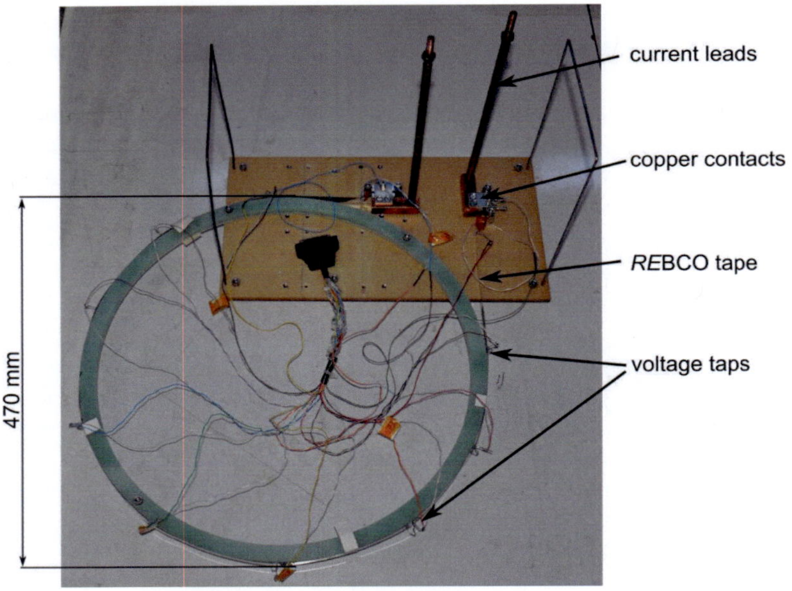

Figure 4.8: Mechanical setup for measuring the critical current at self-field and 77 K. The *RE*BCO tape is bend with the same diameter as in the AC-SFCL.

Figure 4.9 shows a current voltage curve for a 1.475 m long piece of tape A. Section V_{10} shows the lowest critical current of I_c = 266 A. Section V_9 shows the highest critical current of I_c = 272 A.

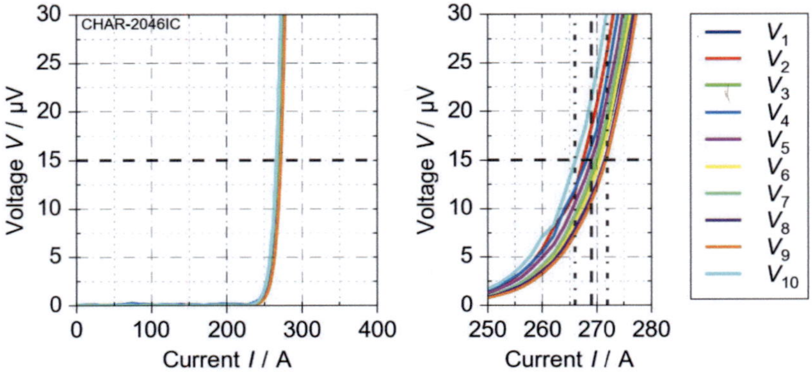

Figure 4.9: Measurement of critical current I_c for a sample of tape A at self-field

The average critical current over the whole tape is I_c = 269 A and the difference between the highest critical current I_c and the lowest critical current I_c is 6 A.

Figure 4.10 shows a current voltage curve for a 1.475 m long piece of tape B. Section V_5 shows the lowest critical current of I_c = 377.5 A, while sections V_4 and V_8 show the lowest critical current of I_c = 381 A. The average critical current along the whole tape is I_c = 379.25 A and the difference between the highest critical current I_c and the lowest critical current is I_c = 3.5 A.

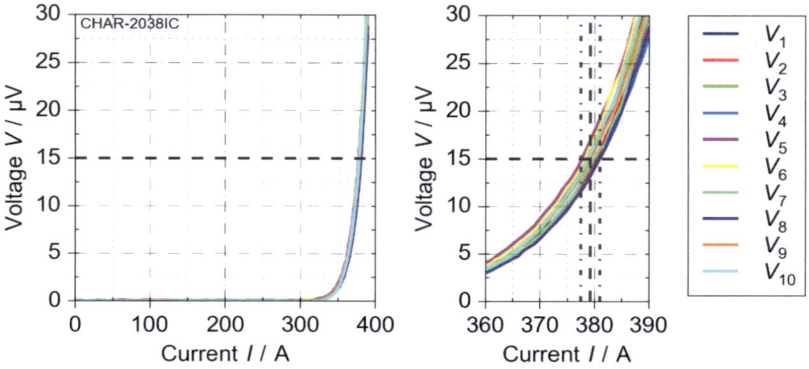

Figure 4.10: Measurement of critical current I_c for a sample of tape B at self-field

The deviations in terms of critical current I_c are below 3% in the examples investigated. This is sufficiently small to ensure, that the tapes will quench completely and instantaneously during fault operation.

4.1.2 Connecting superconducting tapes and quench behavior of a single superconducting loop

To manufacture the rings it was necessary to guarantee a low-ohmic and reproducible soldered connection of *RE*BCO tapes and investigate their quench behavior. Therefore a small test coil was built in which a single *RE*BCO tape was short-circuited and quenched. Additionally, the small coil allowed to examine the manufacturing process itself as well as contacting of voltage taps and measurement setup, which were used in the tests of the demonstrator later on.

Connecting REBCO tapes

A requirement for using *RE*BCO tapes in the AC-SFCL is to manufacture a soldered connection (joint) with low impedance. The soldering process has to ensure a small deviation in of the resistance of all manufactured connections and have to show no degradation of the critical current I_c.

Figure 4.11 shows the principle of the soldered connection. The ends of the *RE*BCO tape loop were put against each other and a tape piece of 6 cm length was used to make the connection with an overlap 3 cm at each tape end. Hereby the *RE*BCO layers of the tapes face each other in order to minimize the distance for the current from *RE*BCO layer to *RE*BCO layer. In former research this type of joint is referred to as bridge joint [MZC15] or butt joint [SD10].

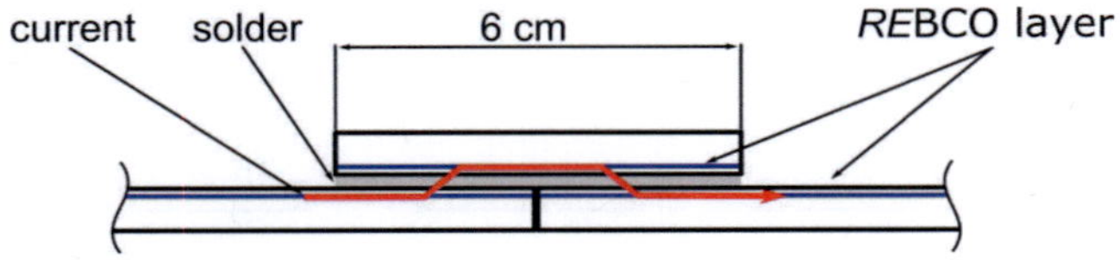

Figure 4.11: Principle of the soldered connection. The ends of a REBCO tape face each other and a second, short *RE*BCO piece is used to make the connection.

To manufacture the soldered connection a fixture was built as shown in Figure 4.12. Firstly the *RE*BCO tape was cleaned with ethanol and pre soldered at the ends on a length of 3 cm using a low temperature solder based on an indium-tin compound. The solder was chosen due to its low melting temperature of T_m = 167 C, which ensured a degradation-free soldering process. The connection piece was pre soldered on the whole surface as well.

A block with a groove of 13 mm width was used to place the *RE*BCO tape inside with the soldered side facing upwards. The connection piece was placed on top the ends and an intender with a fitting notch was used to press the *RE*BCO tapes together. The necessary force was applied using nuts on M4 threads. The nuts were slightly tightened. The whole fixture was then heated to 175°C on a heating plate to liquefy the solder. The nuts were tightened again, this time with 2 Nm torque to form the connection. The heat was applied for two minutes and the fixture then removed from the heating plate and cooled down at room temperature. After cool down the nuts were opened and the soldered connection was removed from the block.

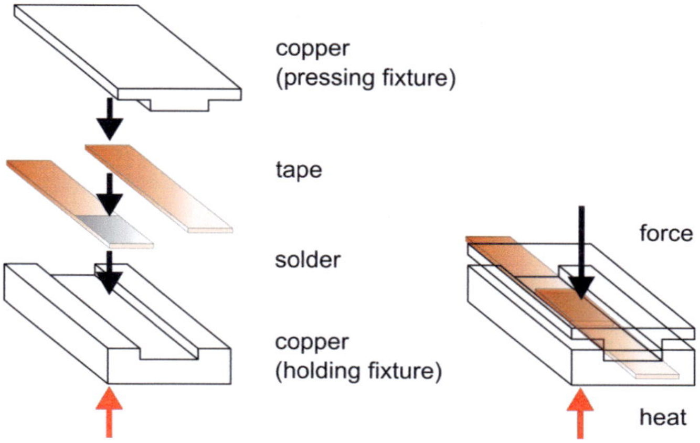

Figure 4.12: Principle of fixture for soldering tapes. The tapes are placed in the holding fixture with the pressing fixture on top. A constant force and heat is applied until the *RE*BCO tapes form the soldered connection.

Before connecting the actual rings, several small samples were manufactured and the resistance measured in order to prove the reliability of the soldering method. As an example Figure 4.13 shows the contact resistance depending on current.

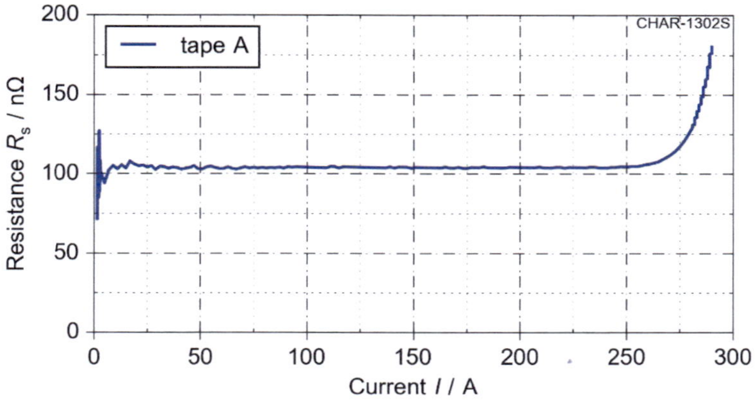

Figure 4.13: Measured resistance depending on current for small soldered pieces of *RE*BCO tapes (6 cm lap length, 12 cm total length)

The setup used for this measurement is a 4 wire sensing based on the setup shown in Figure 4.7. However for this measurement only one voltage drop (across the soldering) was measured

The results show a resistance of R_s = 104 nΩ for a soldered joint as shown in Figure 4.11. The measurement was stopped before the E_c-criterion was exceeded. Referred to the overlap area this is equivalent to R_s = 374.4 nΩcm² for the whole bridge joint or R_s = 187.2 nΩcm² for each lap joint. This result is very well within reported resistances for *RE*BCO joints [PAK07], [BOS09], [LHS11], [MZC15], [BAC15].

Assuming a current of I = 300 A in a *RE*BCO ring with 12 mm width of the secondary winding, the losses P_{sl} due to the soldering would be

$$P_{\text{sl}} = R_{\text{s}} \cdot I^2 = 104\ \text{n}\Omega \cdot (300\ \text{A})^2 = 9.36\ \text{mW} \tag{4.1}$$

A typical secondary winding of an AC-SFCL can contain up to several hundreds of *RE*BCO rings. The losses caused by the soldering in such a winding would remain below 10 W. This is significantly lower than the AC-losses of the conceptual designs presented in chapter 6 for example. Nevertheless the resistance of R_s = 104 nΩ is sufficiently low for the joints in AC-SFCLS demonstrator. Nevertheless, the resistance of REBCO joints for AC-SFCL designs can be further reduced by increasing the overlap length, the use of a solder with lower resistivity and improving the soldering process itself.

Manufacturing the test coil and contacting of voltage taps

In order to investigate the quench behavior of single tapes a small coil was built to test a single tape under fault conditions. The preliminary tests allowed investigation of the manufacturing steps of the primary and secondary winding of the AC-SFCL, the measurement setup for short-circuit tests and contacting of voltage taps. Different *RE*BCO tapes were tested as well as different methods of contacting voltage taps. The measurement setup for the short-circuit tests of a single superconducting loop served as a blueprint for the measurement setup to test the actual AC-SFCL.

The small test coil was made out of G10 with a copper winding as shown in Figure 4.14. In order to quench a single superconducting *RE*BCO ring a mounting was manufactured, which held the ring at its position and offered the possibility to test different methods of attaching voltage taps. A Rogowski coil was used to measure the current inside of the *RE*BCO ring. The tested *RE*BCO ring had a diameter of 249 mm and a respective circumference of 782.3 mm. The main parameters of the test coil are summarized in Table 4.3.

In order to measure the voltage in the short-circuited *RE*BCO ring four kinds of contacting methods have been investigated. The first method was an adoption of the voltage clamps for the I_c-measurement of long tapes (Figure 4.8). Customized stainless steel clamps were used, which pressed themselves on the tape. The signal quality suffered from thermal and mechanical deformation due to cooling and heating during the testing (the tension of the clamps faded) and the influence of the magnetic field of the primary winding. This method of contacting was therefore discarded. Contacts with silver paste turned out to be as prone to mechanical influence due to temperature change and not reliable enough.

Table 4.3: Geometrical and electrical parameters of the fabricated test coil

Parameter	Symbol	Value
Height	h_{tc}	55 mm
Outer diameter	d_{atc}	280.6 mm
Inner diameter	d_{itc}	268 mm
Number of turns	N_{tc}	25
Inductance	L_{tc}	268 mH
Resistance	R_{tc}	30.5 mΩ

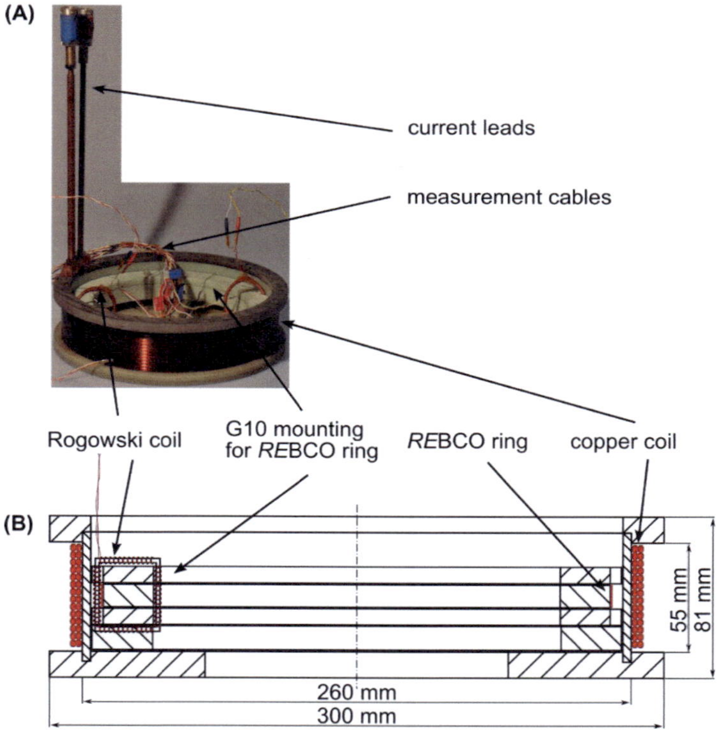

Figure 4.14: (A) Picture of the manufactured test coil and (B) cross-sectional view

Turning the superconducting layer inwards and contacting the tape with gold contacts with springs, which pressed themselves on the tape ensured a continuous and reliable contact during the measurements. However this method introduced a small loop, which in return induced an additional voltage, due to the magnetic field of the primary winding.

The final option tested was soldering the contacts directly onto the *RE*BCO tape. This approach sacrificed the flexibility to change tapes easily and having an additional tension through the gold contacts to keep the *RE*BCO ring in place. On the other hand it allowed placing the measurement cables next to the *RE*BCO tape by machining a small channel in the G10-ring. A superimposed voltage is therefore omitted.

The test coil was connected to a measurement setup shown in Figure 4.15. A 400 V/50 V, 400 kVA transformer was used as power source. Two anti-parallel thyristors allowed triggering and passing of negative and positive half cycles. The resistance R_a was used to adjust the prospective current I_p in the primary winding. Voltages measured were the source voltage V_s, the voltage V_a across the resistance R_a, the voltage V_{tc} at the terminals of the current leads and the voltage V_{sc} across the superconducting *RE*BCO ring. The current in the primary winding I_p was measured using a Rogowski coil. All measured parameters were processed by a transient recorder (TR) and transferred to a computer (PC). The measurement was started and controlled from the PC. A synchronization box (SB) ensured, that the measurement started exactly at the zero-crossing of the voltage of the power source. A protective circuit was installed to cut off any voltages above 9 V in order to prevent damage of the transient recorder [Hie11].

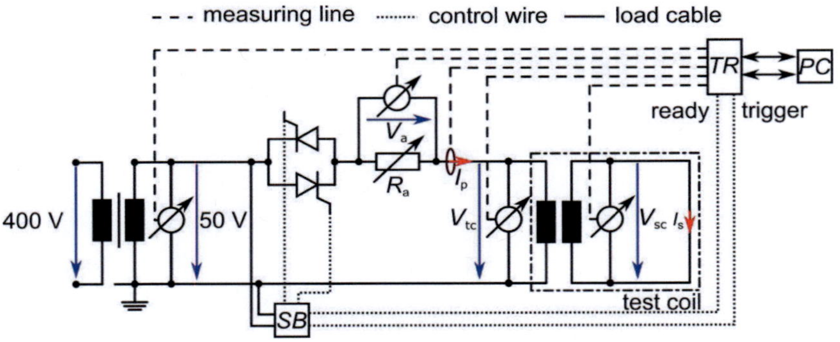

Figure 4.15: Measurement setup for quench experiments of single, short-circuited superconducting *RE*BCO rings

Figure 4.16 shows an example of voltages induced, when a current of I_p = 328 A is conducted in the primary winding of the test coil. The resistance was set to R_a = 150 mΩ and the measured voltage drop at the test coil was V_{tc} = 28.3 V.

The induced voltage in the short-circuited *RE*BCO ring was measured across ten voltage taps as shown in Figure 4.16 on the right. The segments show a very different quench behavior in terms of induced peak voltage. While each segment has the same length of 7.8 cm the voltage peaks differ greatly. The voltage V_8 reaches a peak of 130 mV, which results in an electrical field strength of E_8 = 16.7 mV / cm. The lowest voltage drop was measured at $\widehat{V_5} = 6\,\mathrm{mV}$, this results in an electrical field strength of E_5 = 0.77 mV / cm, which is still above the E_c-criterion of E_c = 1 µV / cm.

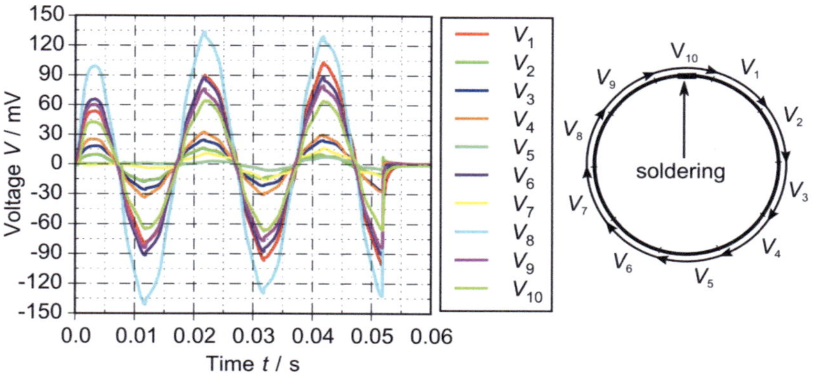

Figure 4.16: Quench results of a single short-circuited *RE*BCO ring (tape B) with ten voltage taps (order is shown on the right)

Comparison with the measurement of the critical current I_c in 10 segments (compare 4.1.1) of the quenched tape showed no particular relation between the distribution of critical current I_c and quench behavior of the respective *RE*BCO tape. This is in contrary to quench measurements of single resistive type tapes and can be explained by the influence of the magnetic field.

4.2 Specification and dimensioning

The purpose of the demonstrator is to confirm the general principle of the AC-SFCL and investigate the operational behavior during normal and fault operation. The equations and the approach described in chapter 3 have been used to design the AC-SFCL demonstrator. Therefore, the AC-SFCL demonstrator verifies the design method as well. For the design and the manufacturing of the AC-SFCL demonstrator the electrical parameters had to be specified and several parameters and constraints must be defined.

Electrical specification

To build the AC-SFCL demonstrator it is necessary to specify the apparent power S_n, voltage V_n, the frequency f_n (50 Hz or 60 Hz) and the reference impedance z_{lim} as described in chapter 3.1. The AC-SFCL demonstrator was intended to operate in a single phase lab environment with the electrical specification summarized in Table 4.4. The specified apparent power S_n and voltage V_n reflect the intended operation. The reference impedance during fault of z_{lim} = 6% is a typical value for air core reactors. Reaching an impedance $z_n \leq 1\%$ at least was considered a reasonable ratio of impedance during fault and impedance during normal operation.

Table 4.4: Electrical specification of the AC-SFCL demonstrator

Parameter	Symbol	Value
Apparent power	S_n	60 kVA
Voltage	V_n	400 V
Frequency	f_n	50 Hz
Reference impedance during fault	z_{lim}	6%
Reference impedance during normal operation	z_n	≤ 1%

From the specified parameters in Table 4.4 the electrical parameters listed in Table 4.5 can be directly derived using eq. (3.9) to eq. (3.12).

Table 4.5: Derived electrical parameters of the AC-SFCL demonstrator

Parameter	Symbol	Value
Rated current	I_n	150 A
Impedance of primary winding	Z_{lim}	160 mΩ
Inductance of primary winding	L_p	0.5093 mH
Impedance during normal operation	Z_n	27 mΩ

The current I_n defines the minimum cross-section A_{Cu} of the conductor in the primary winding. For the primary winding, copper wire with a rectangular cross section and lacquer insulation was chosen. The maximum current density in the copper wire was defined as 5 A/mm². This conservative assumption corresponds with a minimum cross-section of

$$A_{\mathrm{Cu}} \geq \frac{150\ \mathrm{A}}{5\ \frac{\mathrm{A}}{\mathrm{mm}^2}} = 30\ \mathrm{mm}^2 \tag{4.2}$$

Geometry and tape parameters

The demonstrator was designed for operation and testing in a lab environment. Therefore the geometry should be rather compact and the dimensions should not exceed 0.5 m in height and diameter in order to allow easy handling. To facilitate the manufacturing, it was decided to operate the primary winding at 77 K in liquid nitrogen. The primary winding was designed as an air core reactor and the secondary winding as an insert. The parameter range, which was considered to find suitable geometries for the primary winding is given in Table 4.6. Herein the parameter range of inner radius r_{ip} and the height h_p is varied over the desired maximum dimension of 0.5 m. The range of the fill-factor f_{fp} is rather high, but proven to be feasible by the manufacturing of a small test coil.

Table 4.6: Variable geometrical parameters of the primary winding for the AC-SFCL demonstrator

Parameter	Symbol	Value (range)
Inner radius	r_{ip}	0.1 m ... 0.6 m
Height	h_p	0.2 m ... 0.6 m
Fill-factor	f_{fp}	0.8 ... 0.9

With the parameters in Table 4.6 and the electrical specification in Table 4.4 the design process was started in order to find a suitable geometry for the primary winding using the design process described in chapter 3.4. The calculated results for impedance during normal operation depending on number of turns N_p for different geometries or values for inner radius r_{ip} and height h_p are shown in Figure 4.17.

From Figure 4.17 it can be seen, that the minimum inner radius r_{ip} of the primary winding needs to be ≥ 0.2 m in order to reach the specified impedance of z_n = 1%. Figure 4.18 magnifies the region of geometries, which fulfill the z ≤ 1% criterion. Therefore an inner radius of r_{ip} = 0.25 m was chosen, which corresponds to the defined maximum diameter defined for easy handling. The chosen height of the primary winding is h_p = 0.3 m and the number of turns are N_p = 33 in order to maintain a reasonable compact geometry and safely meet the z_n < 1% criterion. For this geometry a secondary winding was calculated as a retrofit according to the design process described in chapter 3.4.1.

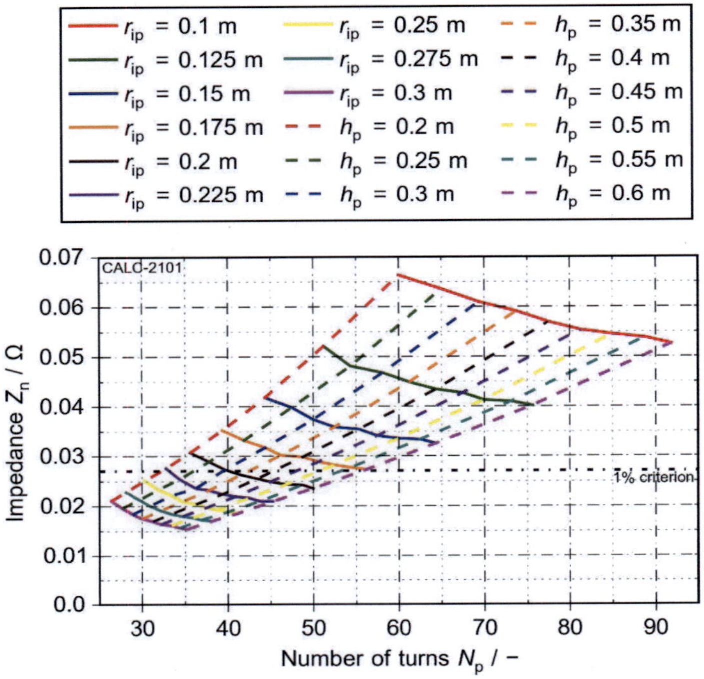

Figure 4.17: Results for impedance during normal operation of the demonstrator in the first design step (fill-factor f_{fp} = 0.85)

For the secondary winding commercially available *RE*BCO tapes with a width of b_{sc}=12 mm with copper stabilization were considered. This minimizes the gaps in axial direction and therefore provides a better shielding as well as the total length of *RE*BCO tape required. The critical current I_c was varied in the given range in Table 4.7 as well as the thickness of the stabilization layer and the winding distance d_w.

Table 4.7: Specification for secondary, superconducting winding and *RE*BCO tapes

Parameter	Symbol	Value (range)
Tape width	b_{sc}	12 mm
Critical current	I_c	200 A ... 400 A
Thickness of Stabilization	h_{stab}	0 µm ...100 µm
Distance of windings	d_w	5 mm ... 20 mm

The parameters given in Table 4.6, Table 4.7 and Table 4.4 were used as input parameters for the design of the retrofit as described in chapter 3.4.1.

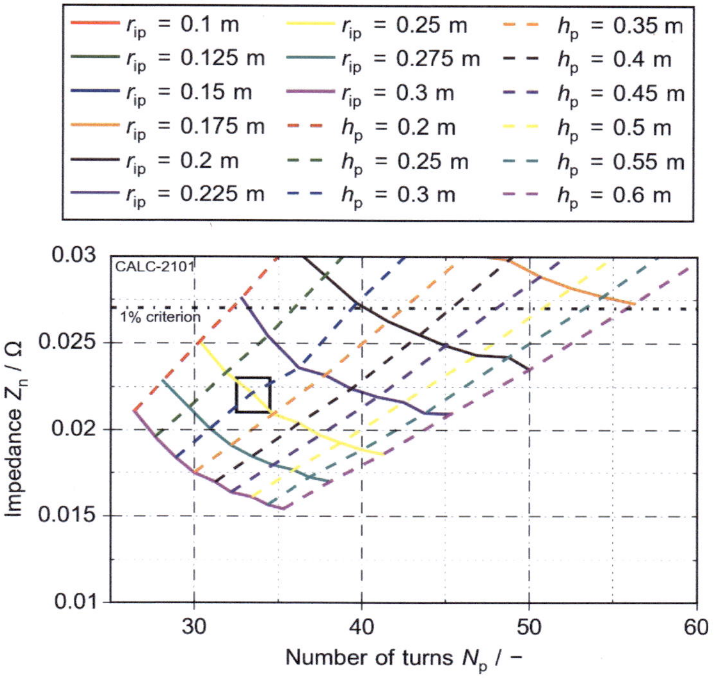

Figure 4.18: Results for impedance during normal operation of the demonstrator in the first design step magnified on geometries fulfill the z ≤ 1% criterion (fill-factor f_{fp} = 0.85). The rectangle indicates the chosen geometry

4.3 Design and expected current limiting capability

During the manufacturing process several improvements could be implemented, which allowed to shrink the geometry even further. The use of PVC reinforcements during the winding process allowed reducing the wall thickness of the bobbin to 4 mm without deformation due to the forces. This reduced the distance of the windings as well to d_w = 6 mm compared to the d_w = 10 mm assumed in the calculation. Subsequently this allowed reducing the overall dimensions of the AC-SFCL demonstrator to an outer diameter d_{ap} = 0.5 m and the height of the primary winding to h_p= 0.275 m. Furthermore the gap between the *RE*BCO tapes in axial direction could be reduced to 0.2 mm resulting in a height of the secondary winding of h_s = 0.266 m. The geometrical parameters of the built AC-SFCL demonstrator are summarized in Table 4.8.

Table 4.8: Geometrical parameters of the AC-SFCL demonstrator

Parameter	Symbol	Value
Total height	h_{demo}	0.34 m
Height of primary winding	h_p	0.275 m
Outer diameter	d_{ap}	0.494 m
Inner diameter primary winding	d_{ip}	0.48 m
Number of turns (primary)	N_p	34
Distance between windings	d_w	6 mm
Diameter secondary winding	d_{is}	0.468 m
Height of secondary winding	h_s	0.268 m
Number of parallel tapes	n_p	22
Gap between tapes	h_g	0.2 mm

Figure 4.19 shows the cross-section of the AC-SFCL demonstrator with indication of the major geometrical parameter. The primary winding was wound using copper wire with rectangular cross section of 8 mm x 3 mm and lacquer insulation. This wire was already used in the small test coil and proofed its temperature resilience at 77 K and allowed high fill-factors of $f_{fp} \geq 0.8$. One turn contains two copper wires resulting in a total cross-section of 48 mm². At the upper and lower end the primary winding is confined by closing rings.

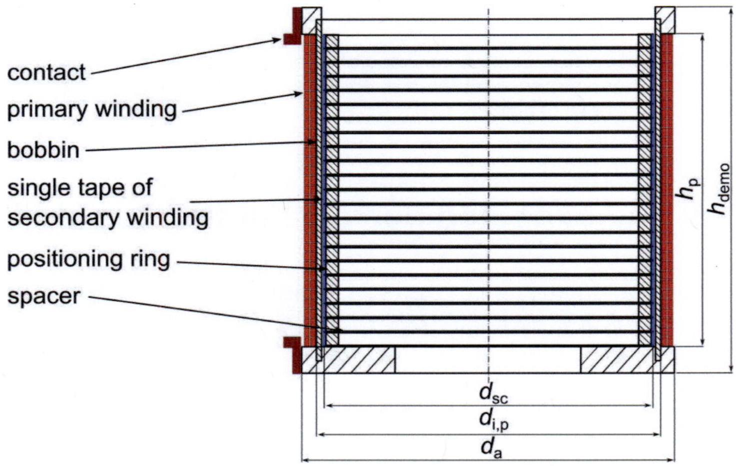

Figure 4.19: Schematic cross-sectional view of the build AC-SFCL demonstrator. The corresponding dimensions are shown in Table 4.8

In the secondary winding each *RE*BCO tape is short-circuited through a soldered connection. Each tape is held at its position by a G10-ring of the same height in parallel.

The tapes are separated by each other by a G10-spacer in axial direction. Each spacer has a thickness of 0.2 mm and is blanked periodically at the outer edge to allow a constant flow of liquid nitrogen (compare Figure 4.21). The secondary winding is assembled by stacking the *RE*BCO tapes with position rings and spacers on top of each other. In total 22 *RE*BCO tapes and 21 spacers are used to reach the height h_p of the primary winding, leaving 3.5 mm of the primary winding at the end and the bottom shielded. The modular design allowed to test different winding configurations and change of *RE*BCO tapes in case of damage.

For the secondary winding of the AC-SFCL demonstrator the *RE*BCO tapes from Superpower (ST12050 [Sup12]) were used. The respective data of the *RE*BCO tapes is given in Table 4.1. Figure 4.20 shows the AC-SFCL demonstrator.

The secondary winding is pulled out. At the outside small screws protrude, which fasten the upper and lower ring to the bobbin. Figure 4.21 shows a close up photo of the secondary winding, exposing the 0.2 mm thick spacer with cuttings for constant flow of liquid nitrogen and the *RE*BCO tapes with position rings. The geometry of the AC-SFCL demonstrator determines the electrical parameters summarized in Table 4.9.

Table 4.9: Electrical parameters of the AC-SFCL demonstrator

Parameter	Symbol	Value
Primary Resistance	R_p	18 mΩ (RT) 2.6 mΩ (77 K)
Primary stray reactance	$X_{\sigma p}$	11 mΩ
Main reactance	X_m	155 mΩ
Secondary stray reactance	$X'_{\sigma p}$	6.9 mΩ
Secondary resistance	R'_s	208 mΩ
Impedance during fault (reference value)	Z_{lim}	(74+110i) mΩ (4.81%)
Impedance during normal operation (reference value)	Z_n	(2.6+18i) mΩ (0.68%)

The resistance R_p of the primary winding could be verified by measuring at room temperature. For the operation at 77 K a RRR = 10 was assumed. The resistance R'_s was calculated assuming the *RE*BCO tapes only generating the smallest resistance measured in normal conducting state (compare chapter 4.1.1).

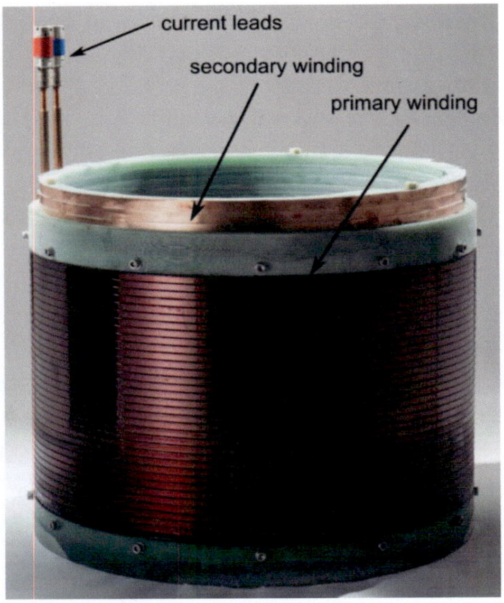

Figure 4.20: Built AC-SFCL demonstrator with secondary winding pulled out

Figure 4.21: Close-up of the secondary winding. The *RE*BCO tapes can be identified by their brushed copper surface. Spacers between the *RE*BCO tapes keep the tapes separated and at position, while the cuttings in the spacers allow the flow of liquid nitrogen.

Expected current limiting capability

The theoretical current limitation was calculated assuming the circuit shown in Figure 4.22. A 400 V power source with an impedance Z_s = 154.7 mΩ or 5.8% was used. This is equal to the transformer impedance used for the measurements described in chapter 4.4. A resistance R_{var} was inserted to adjust the short-circuit current I_s. The AC-

SFCL demonstrator was represented by its resistance R_{SFCL} = 74 mΩ and the reactance X_{SFCL} = i110 mΩ for fault operation. The circuit was as well calculated inserting the resistances R_{SFCL} = 2.6 mΩ and reactance X_{SFCL} = i18 mΩ for normal operation.

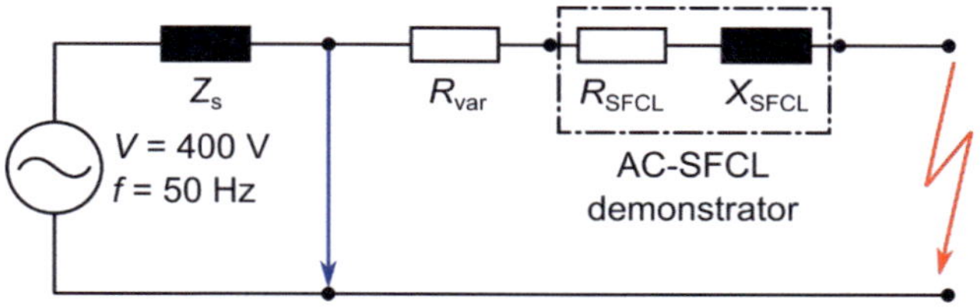

Figure 4.22: Equivalent circuit diagram of the calculated short-circuit

In order to calculate the fault current a one phase steady short-circuit current was assumed and calculated by equations derived from [IEC-60909-0:2001-2007]:

$$i(t) = \kappa \cdot I_k'' \cdot \sin(2\pi f_n \cdot t + \frac{\pi}{2}) \tag{4.3}$$

Herein *i(t)* is the progression of the short-circuit current, *t* the time, f_n the rated frequency. The constant κ is calculated by

$$\kappa = 1.02 + 0.98 \cdot exp(-\frac{3R}{X}) \tag{4.4}$$

with *R* and *X* being the respective values of all resistive and reactive components added up in the short-circuit path. The current I_k'' is calculated by

$$I_k'' = \frac{c \cdot V_n}{Z} \tag{4.5}$$

Wherein V_n is the rated voltage and the constant c = 1 for a branch (far from generator). The impedance *Z* is the total impedance in the short-circuit path:

$$Z = \sqrt{(R_s + R_{var} + R_{sfcl})^2 + (X_s + X_{sfcl})^2} \tag{4.6}$$

Figure 4.23 shows the prospective short-circuit current I_p without AC-SFCL; the current I_{nlim} assuming the impedance during normal operation Z_n is limiting the fault current and the I current I_{lim} assuming the impedance during fault Z_{lim} of the AC-SFCL demonstrator is limiting the fault current.

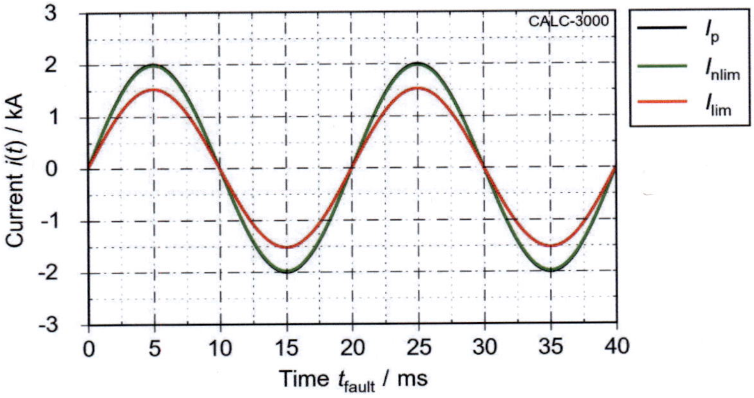

Figure 4.23: Simulation of an expected fault current limitation of the AC-SFCL demonstrator. A prospective short-circuit current of I_p = 2 kA is limited to I_{lim} = 1.5 kA

The calculation shows, that the prospective current of I_p = 2 kA is limited by 25% to I_{lim} = 1.5 kA if with AC-SFCL demonstrator in fault operation. If the impedance during normal operation would be applicable, the fault current would be limited by 1.5% to I_{nlim} = 1.97 A. Figure 4.24 summarizes the simulated results for peak prospective currents up to I_p = 3.5 kA.

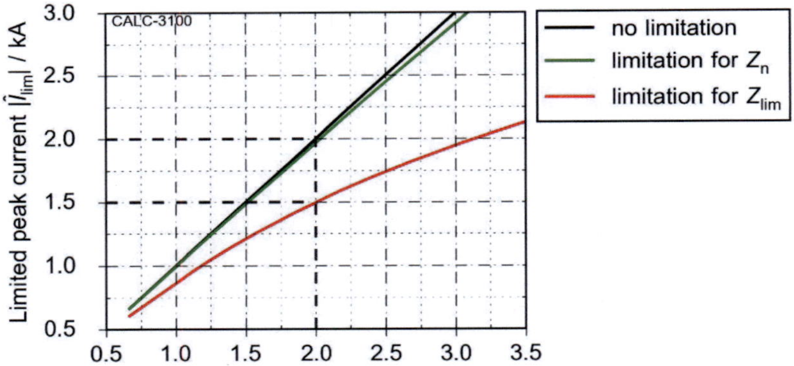

Figure 4.24: Simulation of expected progression of peak limited short-circuit current $|\hat{I}_{lim}|$ with peak prospective current $|I_{p.}|$. The example in Figure 4.23 is indicated.

The simulation shows that the current limitation of the AC-SFCL demonstrator depends on the prospective short-circuit current and increases with increasing peak short-circuit current. While the limitation of the short-circuit current due to the impedance during normal operation is negligible, the impedance of the demonstrator during fault limits the peak short-circuit current by at least 10% starting at I_p = 750 A. This is 5 times the rated current of I_n = 150 A. The peak short-circuiting limitation is increasing with increasing prospective short-circuit current I_p and reaches 39% at I_p = 3.5 kA, which is 23.3 times the rated current I_n. The range of short-circuit currents is identical with the short-circuit range in the actual fault current limitation measurements in chapter 4.4.2.

4.4 Experimental investigation of the operational behavior

Generally, the operation of the AC-SFCL can be categorized in three different operation modes. During normal operation the secondary winding is in superconducting state and fully shields the primary winding, setting the AC-SFCL in low impedance state. During a fault the secondary winding is in normal operation mode, setting the AC-SFCL in high impedance state. After a fault and continued cooling the secondary winding will return to low impedance state. The AC-SFCL is under certain circumstances capable of returning to this state, even while carrying the rated current. This operation mode is called recovery under load. All three operational modes have been investigated.

A 400 V / 400 V transformer was used as power source. The impedance of the transformer was measured as Z_{source} = 154.6 mΩ or z_{source} = 5.8% respectively. The resistance R_p serves as protection of the thyristors and incorporates the resistance of the cables as well and was measured as R_p = 20 mΩ. Two antiparallel thyristors were used as switches for the load branch as well as for the fault branch. The resistance R_l in the load branch was set to 2.67 Ω in order to enable the rated current of I_n = 150 A for normal operation. The resistance R_f was used to adjust the peak current during fault current testing. In both operational modes the current conducted in the primary winding was measured using a Rogowski coil. The measured signal was processed by a transient recorder (TR) as well as the measured voltage drop of the device under test and the source voltage. In addition, voltage taps at 11 *RE*BCO rings in the secondary winding were connected the transient recorder. The transient recorder was controlled via computer (PC). A synchronization box (SB) connected to the source, the thyristors and the transient recorder allowed triggering of load and fault independently at zero-crossing or a chosen phase angle.

For all measurements the AC-SFCL demonstrator was fully emerged in liquid nitrogen at 77 K in an open bath cryostat at normal pressure. Figure 4.25 shows the experimental setup, which was used to investigate all three operation modes.

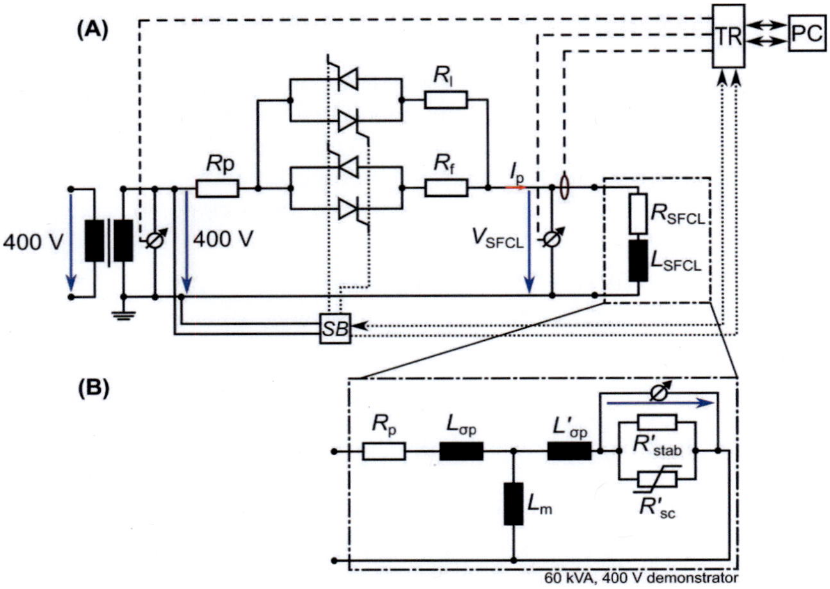

Figure 4.25: (A) Equivalent circuit diagram of the measurement setup used to investigate the AC-SFCLs behavior in all three operational modes and (B) equivalent circuit diagram of the AC-SFCL demonstrator

4.4.1 Normal operation

In order to measure the impedance during normal operation for both, the primary winding only and the AC-SFCL demonstrator, the setup shown in Figure 4.25 was used. Figure 4.26 shows the voltage drop V_{ACR} and the rated current I_n of the primary winding during normal operation at room temperature for a period or four cycles.

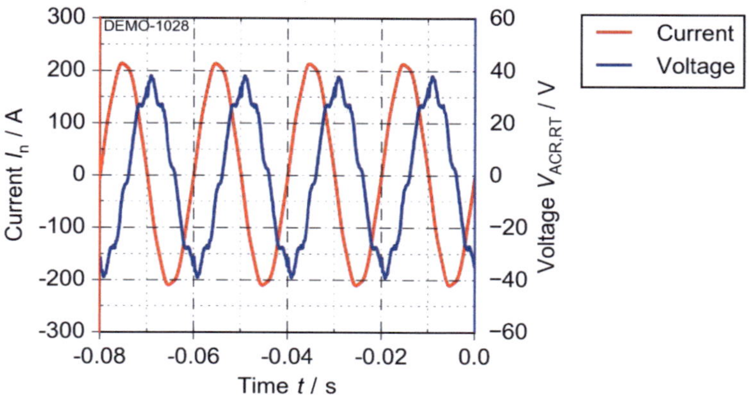

Figure 4.26: Measured current and voltage drop across the primary winding at room temperature during normal operation

The measured peak voltage is $\hat{V}_{ACR,RT}$ = 38.5 V and a current of $\hat{I}_n$ = 212 A. The respective impedance calculates as follows

$$Z_{\mathrm{ACR,RT}} = \frac{27{,}22\ \mathrm{V}}{150\ \mathrm{A}} = 181.5\ \mathrm{m\Omega} \tag{4.7}$$

This corresponds to a reference impedance of z_n = 6.98%.

Figure 4.27 shows the voltage drop V_{ACR} and the rated current I_n of the primary winding fully emerged in liquid nitrogen at 77 K for a period of four cycles. The measured peak voltage is $\hat{V}_{ACR,RT}$ = 38 V and a current of $\hat{I}_n$ = 212 A. The respective impedance calculates as follows

$$Z_{\mathrm{ACR,LN2}} = \frac{26{,}87\ \mathrm{V}}{150\ \mathrm{A}} = 179.2\ \mathrm{m\Omega} \tag{4.8}$$

This corresponds to a reference impedance z_n = 6.89%. The difference in impedance between the primary winding at room temperature and at 77 K is 2.3 mΩ. This impedance change of < 1% is negligible. The measured voltage drop V_{sfcl} at the AC-SFCL demonstrator and the current I_n is shown in Figure 4.28 for a period of four cycles.

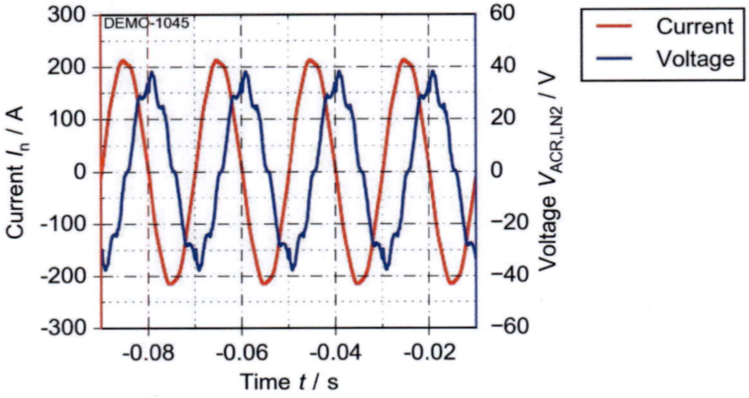

Figure 4.27: Measured current and voltage drop across the primary winding at 77 K during normal operation

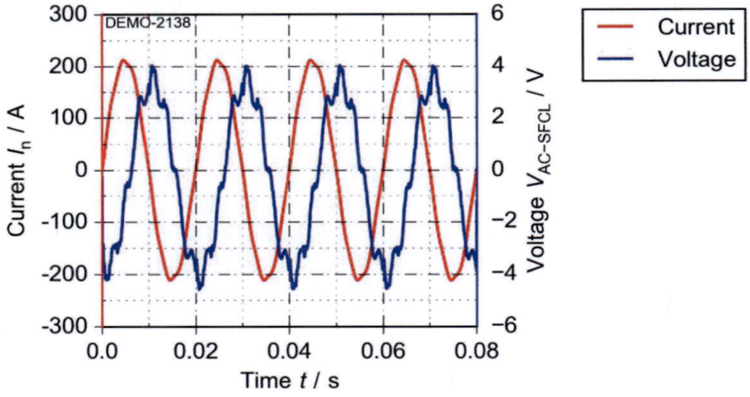

Figure 4.28: Measured current and voltage drop across the AC-SFCL demonstrator during normal operation

Taking into account the offset of V_{off} = 0.4 V (recognizable at the beginning and at the end of the measurement) the measured peak voltage is $\hat{V}_{\mathrm{sfcl}}$ = 4.1 V and a current of $\hat{I}_{\mathrm{n}}$ = 212 A. The respective impedance calculates as follows

$$Z_{\mathrm{sfcl}} = \frac{2.9\ \mathrm{V}}{150\ \mathrm{A}} = 19.3\ \mathrm{m\Omega} \tag{4.9}$$

This corresponds to a reference impedance of z = 0.72%, which fulfills the specification of $z_{\mathrm{n}} \leq 1\%$.

The voltage signals of all three load measurements show distortions and a slight asymmetry. The asymmetry is caused by the magnetization current of the 400 kVA single phase transformer at the zero crossing of the current. The distortions of the voltage signals at their zero crossing is caused by the commutation of the anti-parallel thyristors. The higher frequency distortions are suspected to be caused by the thyristors as well. This suspicion however is still subject to investigation and needs yet clarification.

Table 4.10 summarizes the results for normal operation measurement.

Table 4.10: Comparison of design values and measured impedances for normal operation

	Calculated	Measured
Impedance primary winding at RT ($Z_{ACR,RT}$, $z_{ACR,RT}$)	168.6 mΩ 6.49%	181.5 mΩ 6.98%
Impedance primary winding at 77 K ($Z_{ACR,77K}$, $Z_{ASCR,77K}$)	167.8 mΩ 5.89%	179.2 mΩ 6.89%
Impedance AC-SFCL ($Z_{AC\text{-}SFCL}$, $z_{AC\text{-}SFCL}$)	18 mΩ 0.68%	19.3 mΩ 0.72%

The measured impedance during normal operation z_n = 0.72% of the AC-SFCL demonstrator is in good agreement with the calculated value of z_n = 0.68%. For the primary winding the difference between measurement at room temperature and in liquid nitrogen shows a small difference of 1.3 mΩ.

4.4.2 Current limiting capability during fault

In fault operation the impedance of the AC-SFCL demonstrator increases and effectively limits the fault current. The measurements show the same current limiting capability as the theoretical calculations with the steady-state impedance (chapter 4.3) as well as an increasing current limitation with increasing prospective current I_p.

In order to prove the current limiting capability of the AC-SFCL short-circuit experiments have been performed with prospective fault currents starting at I_p = 650 A up to I_p = 3.6 kA and a duration of up to six half cycles using the experimental setup shown in Figure 4.25.

Figure 4.29 shows the experimental result for a short-circuit of four half cycles and a prospective current of I_p = 2 kA.

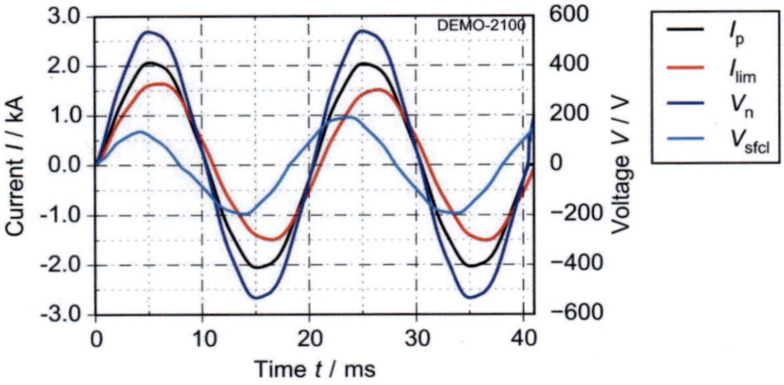

Figure 4.29: Fault current limitation for a fault of four half cycles

The prospective current in the first half cycle is limited from $|I_p|$ =2.06 kA to I_{lim} = 1.64 kA (20.6%), in the second half cycle from $|I_p|$ =2.06 kA to $|I_{lim}|$ = 1.58 kA (23.3%). The fault current limitation of the second cycle is maintained in the third (21.5%) and fourth half cycle (21.5%). The voltage peak in the first cycle is 135.5 V and increases in the second cycle to 196 V, which is maintained in the third and fourth cycle. The voltage drop V_{sfcl} shows a phase shift, indicating an inductive current limitation.

Figure 4.30 summarizes the measured results for fault current limitation for faults with a duration of four cycles for each half-cycle independently.

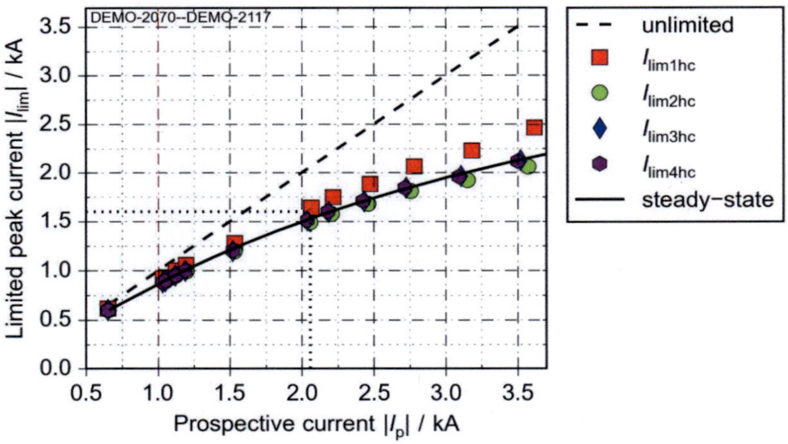

Figure 4.30: Summary of the measured fault current limitation for faults with a duration of four half cycles and the calculated result for steady-state. The case of Figure 4.29 is indicated.

The current limitation in all four cycles depends on the prospective current and increases with increasing prospective current. For prospective currents above $|I_p|$ = 1.5 kA the difference in limitation between the first half-cycle and the other half-cycles becomes noticeable (deviation > 6%). This deviation of the current limitation in the first half cycle becomes more evident with increasing prospective current. The maximum prospective current of $|I_p|$ = 3.6 kA is limited to $|I_{lim}|$ = 2.5 kA in the first half-cycle, this corresponds to 31.2%.In the second half cycle the maximum current limitation is increased to 42.2% and 39.5% in the third and fourth half-cycle. The current limitation of the third and fourth cycle is almost identical with the steady-state simulation. This indicates that the impedance of the AC-AFCL demonstrator is not fully engaged in the first cycle.

In order to compare the measured results with the calculated steady-state values for impedance of the AC-SFCL demonstrator and the air core reactor the reference impedance for each half cycle was calculated individually using

$$z_{\mathrm{hc}} = \frac{z_{\mathrm{lim}}}{Z_{\mathrm{lim}}} \cdot \frac{\hat{V}_{\mathrm{hc}}}{\hat{I}_{\mathrm{hc}}} \tag{4.10}$$

Wherein Z_{lim} and z_{lim} are the specified impedance and reference impedance during fault respectively, V_{hc} is the peak voltage and I_{hc} the limited peak voltage of the respective half cycle.

Figure 4.31 shows the calculated reference impedance of a fault of four half cycles for each half cycle depending on the peak prospective current $|I_p|$ together with the calculated steady-state values.

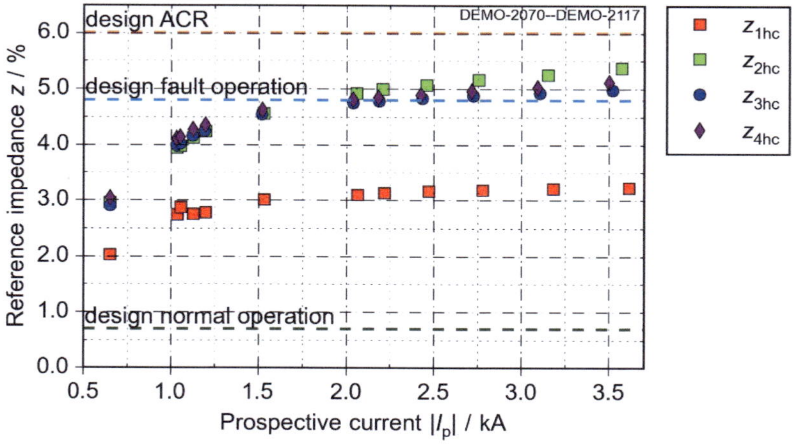

Figure 4.31: Calculated impedance of a fault of four half cycles for each half cycle depending on peak prospective current $|I_p|$. The calculated steady-state values are indicated.

The reference impedance in the first cycle increases from z_{1hc} = 2% close to z_{1hc} = 3% and remains within this range (within < 10%) for increasing peak prospective current $|I_{p.}|$. The following three half cycles show a similar behavior. The reference impedance increases from z = 3% and reaches the calculated reference impedance of z_{lim} = 4.8% at a prospective current above $|I_p|$ = 2 kA. At the maximum peak prospective current the impedance reaches z_{2hc} = 5.37% in the second fault cycle, z_{3hc} =4.98 % and z_{4hc} = 5.11%.

Since the AC-SFCL possesses an inductive and a resistive component, which both limit the fault current. It is therefore necessary to determine the influence of each component. In order to determine the inductive and resistive component of the increased impedance during fault the phase angle φ between limited current and voltage drop over the AC-SFCL demonstrator was calculated. The calculation was executed using the Hilbert transform, which is used for signal processing in Fourier analysis [Foe03], implemented in the scipy software package [Mil11], [Tra07]. The reliability of this approach has been verified in Appendix C.

Figure 4.32 shows the calculated phase angle between limited I_{lim} current and voltage drop V_{sfcl} at the AC-SFCL demonstrator for the fault shown in Figure 4.29.

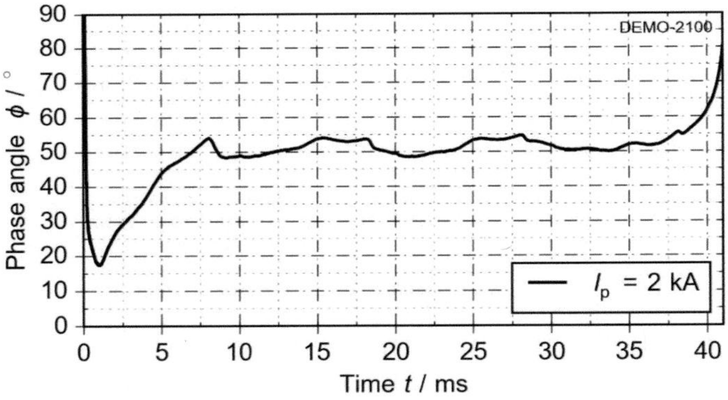

Figure 4.32: Calculated phase angle between current and voltage for the fault of four cycles shown in Figure 4.29

The calculated phase angle φ is increasing from 19 ° to 50 ° for t < 10 ms during the fault. This increase is expected due to the successive breakdown of the shielding and as a consequence the penetration of the magnetic in the air core. As a result, the inductance of the AC-SFCL demonstrator is increasing. For t > 10 ms, after the first half cycle of the fault is passed, the phase angle φ shows a stable progression around 50 ° until the end of the fault at t = 40 ms. The asymptotic progression of the phase angle at the ends of the time interval is caused by the transformation of the signals due to the absence of earlier values. In order to relate a prospective current with a specific phase angle the arithmetic average in the interval 10 ms < t < 40 ms was calculated.

Figure 4.33 summarizes the calculated results of the phase angle φ for the air core reactor at room temperature (RT) and in liquid nitrogen (77 K) and the AC-SFCL demonstrator depending on prospective current $|I_p|$.

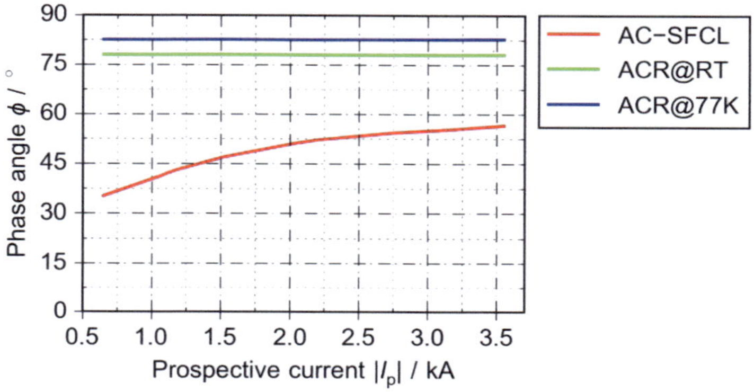

Figure 4.33: Calculated average phase angle of the AC-SFCL and the air core reactor for a fault of four cycles depending on prospective current

The air core reactor shows a constant phase angle at 78° independently of the prospective current I_p. The phase angle increases to 82.6°, if the air core reactor is operated in liquid nitrogen due to the lowered resistance of the copper winding. The phase angle of the AC-SFCL demonstrator depends on the prospective current I_p. It starts at =35.1° and increases to 56.5° in the shown range of the peak prospective current. The transition from mainly resistive to mainly inductive limitation occurs at a prospective current of $|I_p|$ = 1.35 kA.

The knowledge of the phase angle ϕ and the magnitude of the impedance Z_{sfcl} allows the calculation of the resistive R_{sfcl} and the inductive or reactive X_{sfcl} components of impedance [Mar99]:

$$R_{\text{sfcl}} = Z_{\text{sfcl}} \cdot \cos(\phi) \tag{4.11}$$

$$X_{\text{sfcl}} = Z_{\text{sfcl}} \cdot \sin(\phi) \tag{4.12}$$

The calculated (average) resistance R_{sfcl}, reactance X_{sfcl} and impedance Z_{sfcl} for a fault of four half cycles depending on the peak prospective current $|I_p|$ is shown in Figure 4.34.

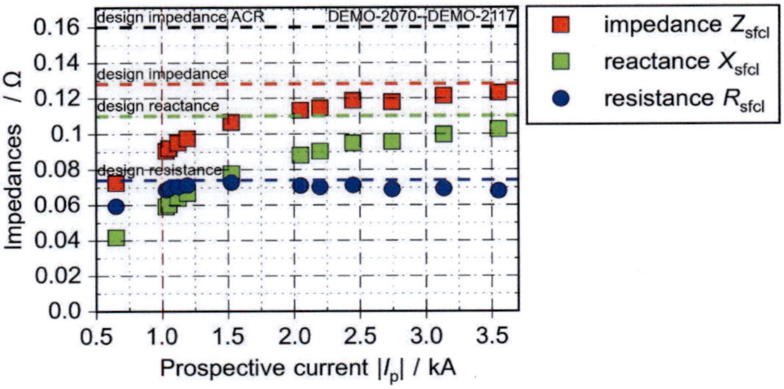

Figure 4.34: Calculated development of resistance, reactance and impedance depending on prospective current

Starting from a peak prospective current of $|I_p|$ = 1 kA the resistance R_{sfcl} remains within 5% of the design value during fault. The reactance X_{sfcl} increases with increasing prospective current and remains below the resistance R_{sfcl} for prospective currents below $|I_p|$ < 1.5 kA. Above prospective currents of $|I_p|$ = 1.5 kA the inductive component dominates the resistive component of the impedance. Both, the reactance X_{sfcl} and the impedance Z_{sfcl} approach their respective design values to the same degree without reaching them in the shown range.

Fault Current Limitation at different phase angle

The presented fault currents all started at the zero crossing of the current. In power systems the fault can happen at any point in time. In order to investigate the fault current limitation of the AC-SFCL a fault of four half cycles with a prospective current $|I_p|$ =2.75 kA was applied and the phase angle varied between 0° and 90°. Figure 4.35 shows a fault for a phase shift of 90°.

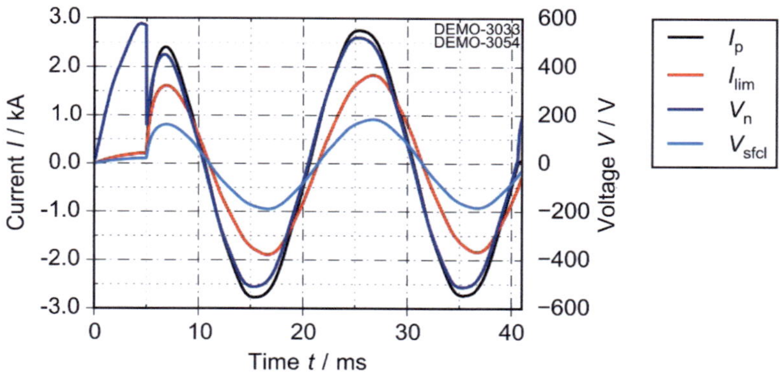

Figure 4.35: Measured limitation of a fault current of four half cycles triggered at a fault angle of γ = 90° and a prospective fault current of 2.5 kA in the first half cycle-.

The prospective current in the first half cycle reaches $|I_p|$ = 2.4 kA and is limited to $|I_{lim}|$ = 1.61 kA, this corresponds to 32.9%. In the following half cycles this current limitation is maintained and corresponds to the current limitation during fault without phase shift. Figure 4.36 summarizes the current limitation in the first peak depending on phase angle γ.

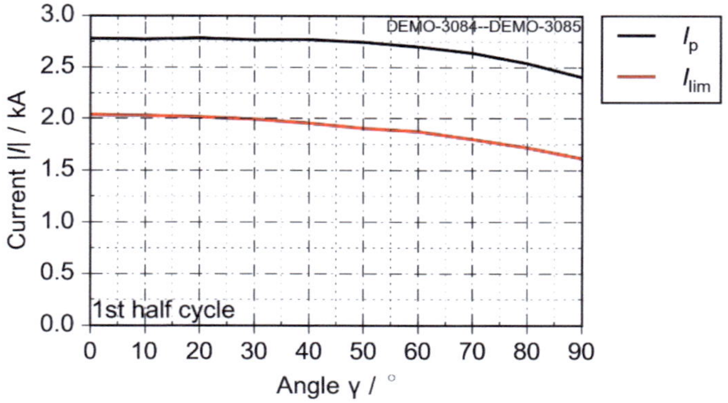

Figure 4.36: Measured fault limitation capability for a fault angles between 0° and 90° in the first half cycle

The prospective current and the limited peak decrease with increasing phase angle γ. Since both currents qualitatively decrease to the same degree an effective current limitation is maintained for all phase angles. The current limitation capability of the AC-SFCL demonstrator for all four half cycles depending on phase angle is summarized in Figure 4.37.

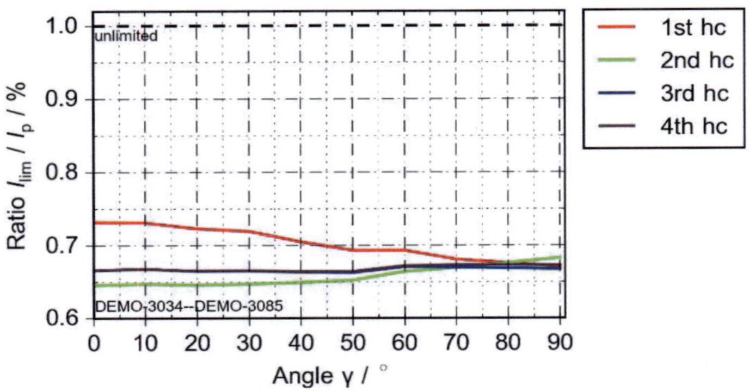

Figure 4.37: Measured fault current limitation for a fault angles γ between 0° and 90° for all four half cycles

For the first half cycle the current limitation is increasing form 26.8% at $\gamma = 0°$ to 33% at $\gamma = 90°$. In the second half cycle the current limitation is slightly decreasing from 35.5% to 31.7%, while it remains unchanged for the third and fourth cycle. The current limiting capability of the AC-SFCL demonstrator is independent of the phase angle of the fault.

4.4.3 Investigation of quench behavior and recovery under load

To achieve this the superconducting tapes must recool below their critical temperature T_c under load. To investigate the recovery under load capability and the quench behavior of the AC-SFCL demonstrator, load cycles were applied before and after the fault and the voltage drop of the *RE*BCO tapes in the secondary winding was measured. Figure 4.38 shows a fault of two half cycles with a prospective current of $|I_p|$ =3.6 kA.

The prospective current in the first cycle is limited to $|I_{lim}|$ =2.45 kA in the first cycle (31.9%) and to $|I_{lim}|$ = 2.05kA in the second cycle (43%). This is in accordance with the current limitation shown in Figure 4.30. The voltage drop measured at the AC-SFCL demonstrator shows the expected phase angle between voltage and current, indicating inductive limitation.

The voltage was measured at 11 of the 22 *RE*BCO tapes of the secondary winding. These 11 *RE*BCO tapes are located in the upper half of the AC-SFCL demonstrator. This is sufficient for two reasons: Firstly, the distribution of the magnetic field in the AC-SFCL demonstrator is symmetrical in axial direction. Secondly, the *RE*BCO tapes were placed in a symmetrical manner as well. The critical current I_c of a *RE*BCO tape in the upper half is similar to the critical current I_c of the *RE*BCO tape at the same position in the lower half.

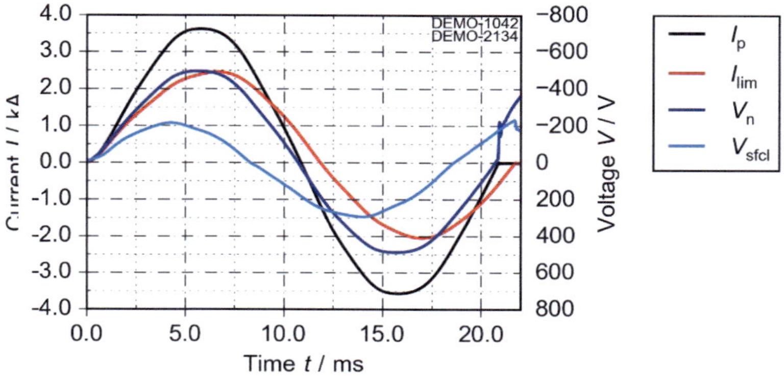

Figure 4.38: Fault current limitation for a fault of $|I_p|$ = 3.6 kA and a duration of two half cycles

The voltages measured at the *RE*BCO tapes for the fault shown in Figure 4.38 are shown in Figure 4.39.

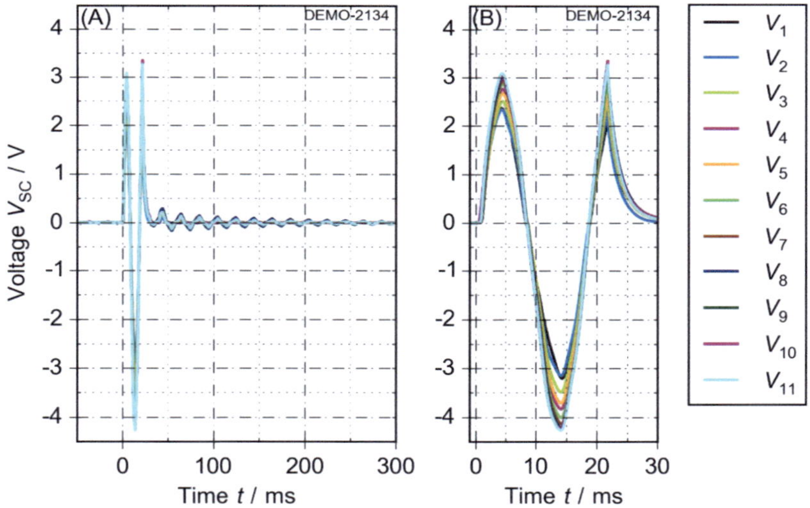

Figure 4.39: Voltage signals at the *RE*BCO tapes during fault current limitation for a fault of two half cycles (A) including the load cycle and (B) magnified on the fault.

The voltage is shown for 11 *RE*BCO tapes starting at the upper end (V_1) of the AC-SFCL demonstrator to the middle (V_{11}). In the first half cycle of the fault the voltages peak at

3.04 V maximum and 2.33 V minimum. In the second half cycle the voltages peak at -4.24 V maximum and -3.17 V minimum, when the fault is switched off.

Table 4.11 summarizes the measured voltages of the *RE*BCO tapes from the outermost tape (No 1) to the center tape (No. 11). At the first seven positions *RE*BCO rings made of tape B with 100 µm copper stabilization were used and for the innermost three positions *RE*BCO rings made of tape A with 40 µm copper stabilization. Using the measured *R*(*T*) values of the *RE*BCO tapes (chapter 4.1.1) the peak current for each *RE*BCO tape in each half cycle is calculated.

Table 4.11: Measured voltages and calculated currents of the tapes during fault

Tape Position No.	U_{1hc} / V	U_{2hc} / V	I_{1hc} / kA	I_{2hc} / kA	average I_c / A (chapter 4.1.1)	type
1	2.33	-3.17	2102	-2849	379	tape B
2	2.30	-3.12	2073	-2797	343.5	tape B
3	2.47	-3.46	2223	-3103	358.5	tape B
4	2.72	-3.80	2453	-3402	351.5	tape B
5	2.63	-3.68	2368	-3297	356	tape B
6	2.83	-3.98	2548	-3558	360.5	tape B
7	2.89	-4.10	2604	-3669	354.5	tape B
8	2.95	-4.17	2662	-3737	363.5	tape B
9	2.99	-4.18	1172	-1627	268	tape A
10	2.98	-4.20	1168	-1635	265.5	tape A
11	3.04	-4.24	1194	-1650	271.2	tape A

The measured peak voltages in each cycle show a different peak voltage depending on the position. They show the tendency to increase from the end towards the center of the secondary winding, with only the voltage of tape 5 stepping out of line. Comparing the peak voltages between the half cycles, the magnitudes in the second half cycle are higher than in first half cycle. This corresponds to the observation that the impedance in the first half cycle is not fully engaged and the maximum current limitation is not reached before second half cycle. The innermost three *RE*BCO tapes (No. 9, 10, 11) show a significant lower current in both half cycles compared to the other *RE*BCO tapes. This is a result of the thinner copper stabilization of tape A (40 µm) compared to tape B with (100 µm) copper stabilization, which generates a higher resistance in normal operation.

In order to investigate the quench behavior in more detail the voltages of the *RE*BCO tapes at the beginning of the fault are shown in Figure 4.40.

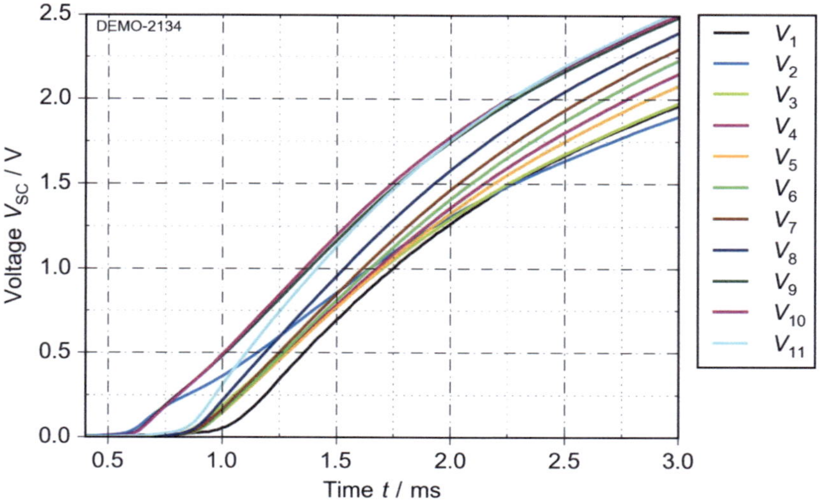

Figure 4.40: Magnification of the voltages at the *RE*BCO tapes in the first half cycle, after engaging the fault

The first quench starts at t = 0.6 ms and within less than 1 ms all tapes started to quench. The tapes quenching first, roughly 0.3 ms before the other tapes, are tape 10, 9 and. Except tape 1, all other tapes are quenching within a time window of $t \leq 0.1$ ms. Around t = 1.5 ms almost all voltages are in an order they maintain for the rest of the shown time frame of the fault. The reason for this behavior is, that all the *RE*BCO tapes in the secondary winding are connected in parallel. The total induced current in the secondary winding is distributed between the *RE*BCO tapes. Whenever the induced current in a *RE*BCO tape exceeds the critical current I_c of the tape it starts to quench. This explains, why the *RE*BCO tapes with lower critical current I_c quench first. An exception to this observation are voltage V_5 and voltage V_2. While the progression of the voltage is consistent with its peak value compared to the other voltages, the progression of voltage V_2 does not follow the same slope as all other voltages. A reasonable suspicion would be, that the measurement cables are affected by the magnetic field of the demonstrator inducing a voltage, which superposes the measured voltage signal.

To determine the recovery time the voltage of each *RE*BCO tape is measured in the following load cycle. To decide, if a *RE*BOC tape is in superconducting state or in normal state a threshold voltage V_{thr} is defined. The threshold voltage takes into account the average offset voltage of the tape measured in the load cycle before the fault and a safety margin for noise. Adding up these voltages lead to a threshold voltage of $|V_{thr}| \leq 2$ mV. A *RE*BCO tape is considered recovered when no peaks occur above the defined threshold voltage V_{thr} for the duration of one half cycle.

Figure 4.41 shows a magnification of the voltage drop of two different *RE*BCO tapes before, during and after the fault with indication of the voltage peaks above V_{thr}. The fault applied is the same as shown in Figure 4.38: Duration of two half cycles and prospective current of $|I_p| = 3.6$ kA.

The voltage V_3 in Figure 4.41 (A) shows declining peaks after switch-off of the fault and switch-on of the load at $t = 20$ ms. The magnitudes of the voltages peaks remain above V_{thr} until $t = 134$ ms. After this point in time no more peaks are detected above V_{thr} and the tape 3 is considered to be in superconducting state at $t = 144$ ms. This means, that the recovery time of tape 3 is $t_{rec} = 124$ ms, the difference between the return to superconducting state and the switch-off of the fault at $t = 20$ ms. The voltage V_{11} in Figure 4.41 (B) shows declining peaks of the measured voltages after the fault is switched-off and the load is switched-on. For $t > 200$ ms the voltage signal V_{11} shows constant peaks with a magnitude of 20 mV until the load is switched-off as well at $t = 510$ ms. This means the voltage in Figure 4.41 (B) shows no recovery.

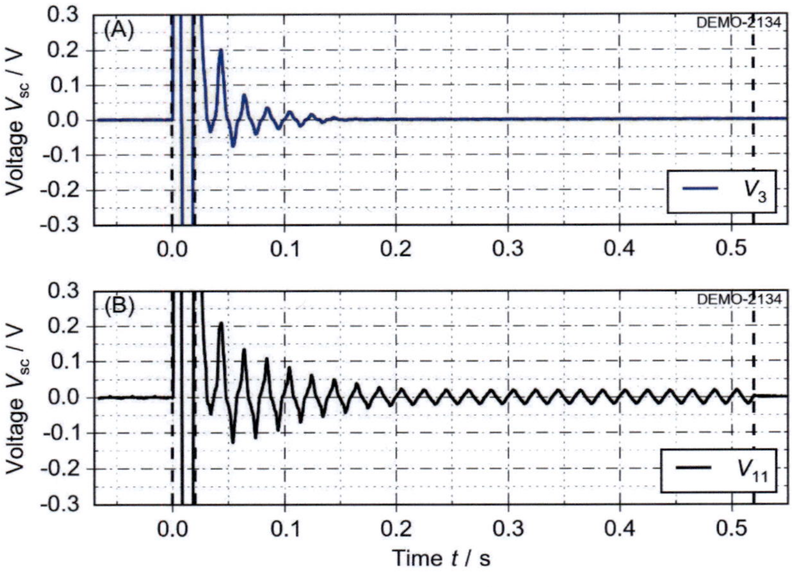

Figure 4.41: Voltages of two *RE*BCO tapes during fault and following load cycle. (A) shows the recovery of tape 3, which recovers 124 ms after the fault is switched of, while in case (B) the tape is not recovering

The complete secondary winding is considered in superconducting state, when all tape voltages returned to superconducting state after the fault is switched off and a load cycle follows. Figure 4.42 summarizes the results for recovery under load for short-circuits with different prospective current and a duration of two half cycles.

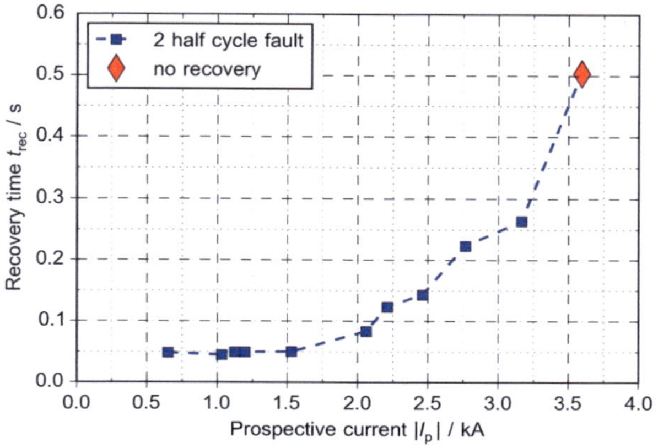

Figure 4.42: Summarized results of recovery under load. At $|I_p|$ = 3.6 kA the tape shown in Figure 4.41 is not recovering

For prospective currents below $|I_p|$ = 1.5 kA the recovery time t_{rec} is nearly constant and remains below 60 ms. For prospective currents higher than $|I_p|$ =1.5 kA the recovery time t_{rec} starts to increase. This increase is corresponding with results of former measurements [BNK11].

4.5 Summary

This chapter describes the work done to build and test an AC-SFCL demonstrator. The manufacturing process, measuring and quenching of single *RE*BCO rings in a small coil were investigated and the practical experience gained could successfully applied to the AC-SFCL demonstrator. Two different *RE*BCO tapes were characterized and tested for their suitability for the secondary winding. The measurement of the resistance depending on temperature $R(T)$ allowed to calculate the steady-state impedance and predict the current limitation of the AC-SFCL demonstrator. The dependency on magnetic field of the critical current $I_c(B,T\text{=}77\text{ K})$ of the used *RE*BCO tape showed a sufficient current carrying capability for use in the AC-SFCL demonstrator. The measurement of critical current I_c in self-field of each *RE*BCO tape piece verified its suitability for the AC-SFCL demonstrator.

The AC-SFCL demonstrator has been designed using the design method and equations presented in chapter 3. It was specified for 400 V, 60 kVA with an impedance during fault of z_{lim} = 6% and an impedance of z_n < 1% during normal operation. The approach hereby was to design the primary winding as air core reactor and retrofit it with the secondary superconducting winding. Reinforcements applied during the winding process allowed to shrink the overall dimensions without violating the specification, primarily the specified values for impedance during normal operation and fault operation.

The following conclusions can be drawn from the performed tests:

- The load measurements proved the calculated steady-state impedance of the AC-SFCL demonstrator (calculated: z_n = 0.68%, measured: z_n = 0.72%) and the impedance of the primary winding (calculated: $z_{n,ACR}$ = 6.49%, measured: $z_{n,ACR}$ =6.89% at 77 K), which is summarized in Table 4.10 and shown in Figure 4.17 to Figure 4.19.
- The highest measured fault current limitation was 31.2% in the first half cycle, 42.2% in the second half cycle and 39.5% in the third and fourth half cycle at a prospective current of I_p = 3.6 kA, which corresponds to $17 \cdot I_n$. (shown in Figure 4.29).
- The AC-SFCL showed increasing and effective current limitation with increasing prospective current. The current limitation increased as well in the second half cycle of the fault and showed good agreement with simulations using the steady-state values for impedance in fault condition (shown in Figure 4.29).

Finally, these results confirm the concept of the AC-SFCL and the design parameters. Furthermore calculation of the impedance for each half-cycle confirmed, that the AC-SFCL demonstrator generates its full impedance in the secondary cycle and maintains it in the following cycles for prospective currents above I_p > 1.5kA. Further investigation showed, that the current limitation of the AC-SFCL demonstrator becomes more inductive than resistive at prospective currents of I_p > 1.35 kA and the resistance reaches its design value (within 5%) at I_p > 1 kA and maintains this resistance for increasing prospective current, while the reactance increases further. Variation of the phase angle of the fault, showed, that the AC-SFCL demonstrator is capable of maintaining its fault current limitation independently of the phase shift.

Investigation of the quench behavior showed, that REBCO tapes with lower critical current I_c (approx. 260 A vs. 360 A in average) are quenching first in the stacked structure of the secondary winding. Recovery under load measurements showed, that the AC-SFCL demonstrator was capable of recooling under load within 300 ms for fault currents, which are $I_f = 15 \cdot I_n$.

5 Power Hardware-in-the-Loop test of the Air Coil Superconducting Fault Current Limiter Demonstrator

Power Hardware in-the-Loop testing (PHIL testing) is an approach to combine experimental testing with computational simulation in order to investigate and analyze the behavior of electrical devices under realistic test conditions [JGV11] and was successfully used for testing SFCLs in the past [SLS09], [DSB11], [GCK14].The principle of a PHIL setup is shown in Figure 5.1.

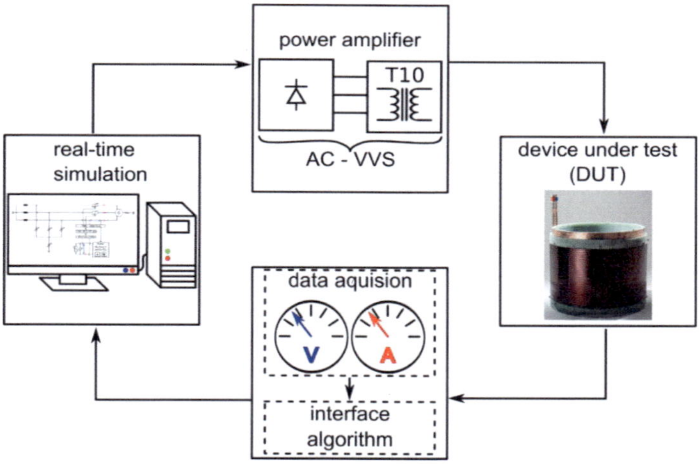

Figure 5.1: Principle of the power hardware in-the-loop setup

In a PHIL system the device under test (DUT), e.g. a SFCL, is physically connected to the power hardware, which usually consists of transformers and power switches to deliver the voltage and current within a certain power range. The power hardware receives control signals from the simulated real-time environment or real-time simulator. This simulated environment can contain any electrical devices, such as transformers, generators, cables or loads, which are implemented as transient models.

The power hardware is responsible for delivering currents and voltages for the device under test. The following tests and experiments were performed at the Center for Advanced Power Systems (CAPS) at the Florida State University (FSU). The test bed at CAPS is capable of delivering 5 MW, 4.16 kV through a variable voltage source (VVS), which is a three phase switching power amplifier [SES10]. The real time simulator, responsible for the simulated environment, supports execution of electromagnetic transient simulations in real-time. The time-step of the real-time simulations was set to 50 µs.

A key role in a PHIL system is the interface algorithm or interface method. It represents the device under test in the simulated environment of the PHIL system. It is therefore necessary to implement an interface algorithm, which delivers reliable results in terms of accuracy, while maintaining a stable operation of the whole PHIL system [RSB08], [PE13]. For the test with the AC-SFCL three interface methods have been investigated and the most suitable was implemented in the 3-phase testing.

5.1 Test validation

Before setting up the PHIL system with the hardware and simulated components it is necessary to test and verify the stability and reliability of the setup as well as preventing critical fail states of the system, which could damage the AC-SFCL or other hardware components. Therefore a simulated model of the AC-SFCL was implemented in the software part of the PHIL system including the circuit of the measurement setup at KIT (compare Figure 4.25). The simulated model of the AC-SFCL demonstrator included all resistances and inductances of the equivalent circuit diagram (Figure 3.3). The *RE*BCO in the secondary winding was modeled as a switch, which would trigger the design resistance, if the current exceeds the critical current. The results for fault current limitation were verified with the results of the measurements (compare chapter 4.4.2).

Figure 5.2 shows the current in the AC-SFCL demonstrator and the voltage drop during a fault with a prospective current I_p = 2 kA and a duration of four half-cycles for the computational model and the measurement performed at KIT.

In terms of fault current limitation the measurement and the computational model are in good agreement (deviations ≤ 2%) for the prospective currents $I_{p,meas}$ and $I_{p,sim}$ as well as for the limited currents $I_{lim,meas}$ and $I_{lim,sim}$. The voltages $V_{sfcl,meas}$ and $V_{fscl,sim}$ differ slightly. The reason therefore is, that the switches in the computational model to trigger the fault current and load respectively (compare Figure 4.25) reproduced the functionality of the thyristors of the measurement setup, but not the electrical behavior and properties. The results for fault current limitation for different fault current peaks are summarized in Figure 5.3.

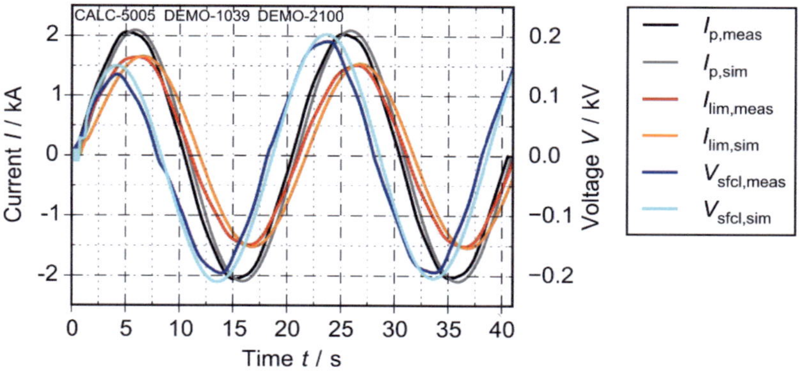

Figure 5.2: Comparison of the simulated model and measured results at KIT for fault current limitation for a fault current with I_p = 2 kA and a duration of four half-cycles

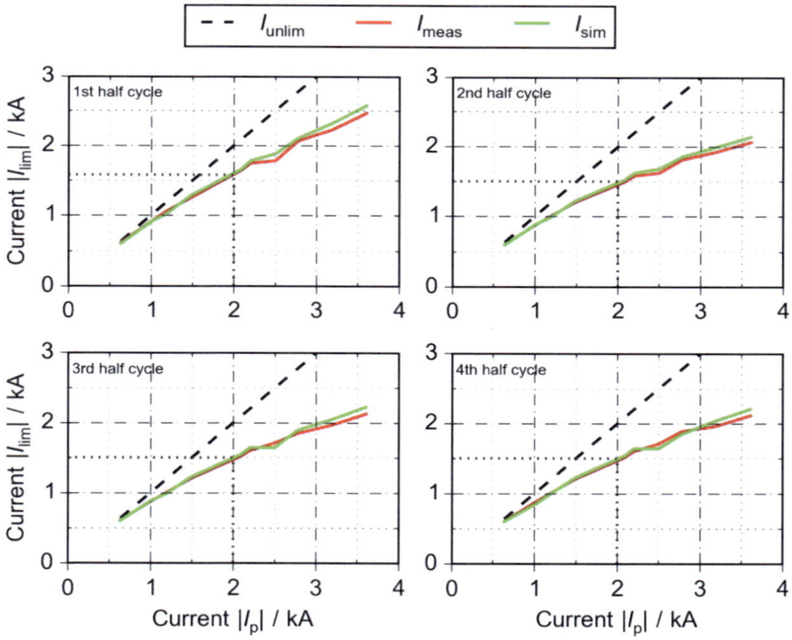

Figure 5.3: Comparison of the simulated model and measured results at KIT for fault current limitation for a fault currents with increasing prospective current (the example in Figure 5.2 is indicated)

The computational model shows the same behavior in terms of fault current limitation as the measurements. The fault current limitation is increasing with increasing fault peak current. For the peak current range shown in Figure 5.3 the measured and computed results are in good agreement (deviations ≤ 4%).

After the validation of the computational AC-SFCL model it was implemented in a simulated setup of the actual PHIL-test setup. In this setup it was assumed, that the current in the AC-SFCL and the voltage drop across the AC-SFCL are fed back into the simulated system. This approach is known as the ideal transformer method (ITM) [Ren08]. During these preliminary investigations with the computational model of the AC-SFCL, it was found, that the ideal transformer method as interface algorithm failed to maintain stability and was therefore discarded for the foreseen tests.

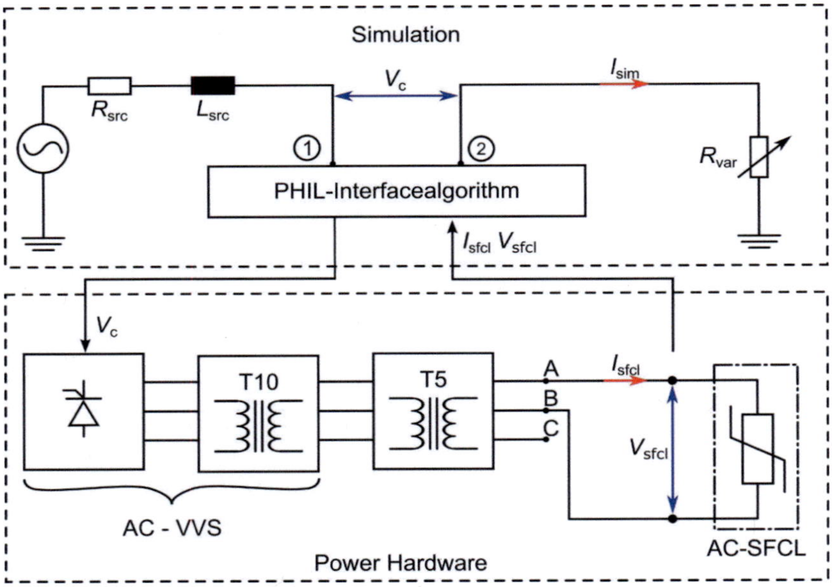

Figure 5.4: PHIL setup for verification of results and testing of different interface methods. The AC – VVS is described in [SES10].

As an alternative three other options for interface algorithms were investigated. These interface algorithms are variants of the damping impedance feedback method [Par13], [Dmi96]. The details of the interface algorithms and the measurement results are given in chapter 5.2. For the verification of the results the modified damping impedance method (mdim) was chosen. The circuit diagram of the PHIL setup used to verify the obtained results is shown in Figure 5.4.

Herein the resistance R_{src} and the reactance X_{src} represent the impedance of the power source (400 kVA/400 V transformer) used at KIT. The resistance R_{var} is used to adjust the current amplitude. The AC-SFCL demonstrator is connected between the terminals A and B of the transformer T5 and fully submerged in LN_2 in an open bath cryostat at normal pressure. The transformer T5 provides the necessary 400 V and is connected to the AC-VVS (variable voltage source), which generates the output corresponding to the control signal V_c from the simulated system.
Figure 5.5 shows the result for load operation and fault current limitation for a prospective current of I_p = 2 kA for four half cycles using the modified damping impedance method described in chapter 5.2 as interface algorithm.

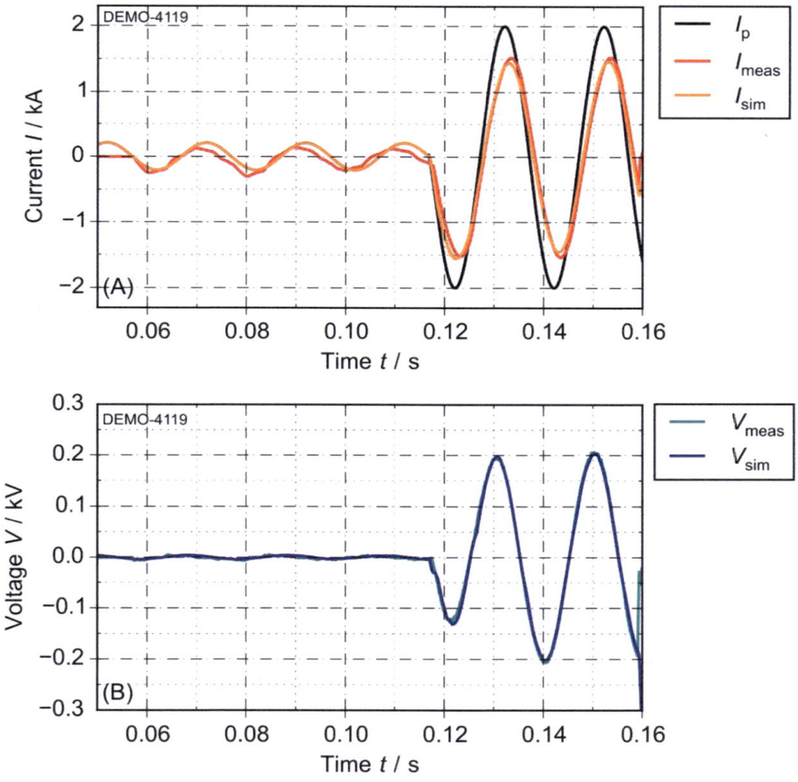

Figure 5.5: Comparison of results for three cycles load operation and limitation of a short-circuit current of I_p = 2 kA with a duration of two cycles: (A) currents and (B) voltages

The measurement is initiated with a load with a duration of six half cycles to minimize the influence of possible transients. In case of simulation of the complete setup the fault current is limited to I_{sim} = 1.54 kA in the first half cycle and I_{sim} = 1.46 kA in the other half cycles. For the measurement the fault current is limited to I_{meas} = 1.52 kA in all four half cycles. The measured voltage V_{meas} and the simulated voltage V_{sim} match each other almost exactly.

The results of the measurement and the simulation are in good agreement for all load cycles and for the four half cycles of fault operation with deviations below 4% in terms of limited peak currents.

Figure 5.6 shows the generated resistance R_{sfcl} and reactance X_{sfcl} as well as the calculated impedance Z_{sfcl} during the measurement.

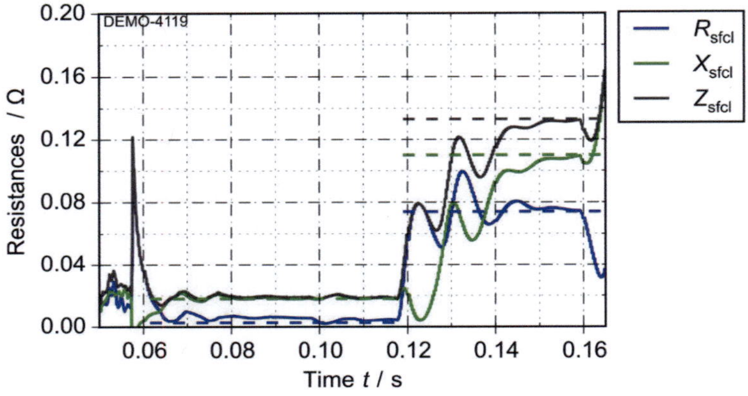

Figure 5.6: Calculated values for resistance, reactance and impedance during the load and the fault cycle. The respective design values for load and fault are plotted as dashed lines.

A short transient oscillation occurs after switching on the load, due to the lack of foregoing values for the calculation. Then the calculated load resistance stabilizes with an offset of 2 mΩ compared to the design value (R_n = 2.6 mΩ), while the reactance matches its design value of X_n = 18 mΩ very well. The resistance offset is introduced by the placement of the voltage taps outside of the LN_2 bath. After the fault is switched on, transients occur in the first cycle due to the windowing of the discrete Fourier transformation. This settles in the second cycle of the fault and the instantaneous calculated values for resistance and reactance converge with the design values.

The results for current and voltage using the modified damping impedance are consistent for PHIL simulation and PHIL measurement. This measurement was repeated with different peak fault currents and compared to reference measurements and calculations from KIT in order to eventually verify the PHIL setup. The results are summarized and compared with previous KIT results in Figure 5.7, which shows

prospective current I_p versus the limited current I_{lim} for each half cycle for a fault duration of four cycles.
The measurements from KIT are described in chapter 4.4.2, the calculation in chapter 4.3, eq. (4.2) through eq. (4.5). The prospective current range for the PHIL tests was defined between 0.5 kA minimum and 2.5 kA maximum in order to keep a good safety margin and prevent fault currents that could damage the AC-SFCL demonstrator.

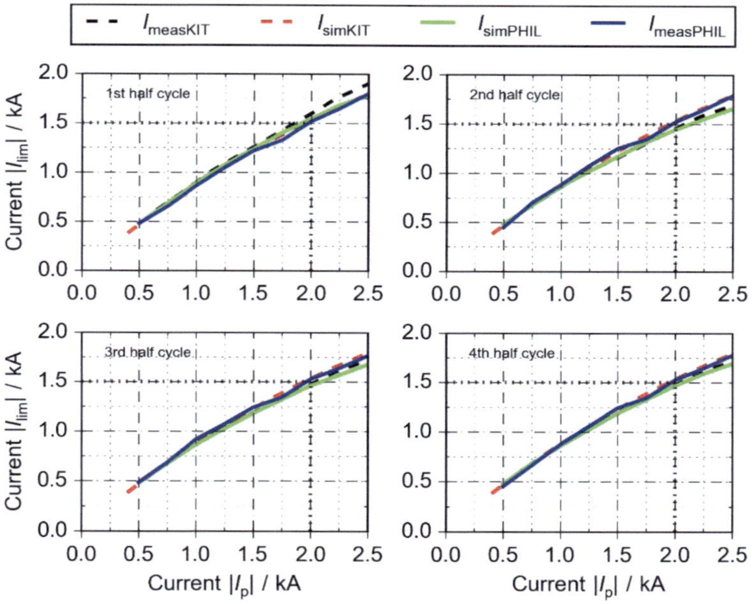

Figure 5.7: Comparison of PHIL results with KIT results for fault current limitation during four half cycles for increasing prospective currents. The example shown in Figure 5.5 is indicated.

In general the results for all four half cycles are in good agreement with deviations < 5%. The biggest difference between the measurements is found in the first cycle. The choice of interface algorithm and the PHIL setup can therefore be considered suitable for the 3-phase testing and the results can be considered reliable.

5.2 Comparison of different interface methods

The interface algorithm is a crucial part of the PHIL system. It is the representation of the physical device under test in the simulated environment. It is therefore important to find an interface method, which ensures a stable operation, reliability of results and accuracy. For the PHIL experiments of the AC-SFCL demonstrator three interface algorithms have been investigated:

- Impedance feedback method (IFM)
- Classical damping impedance method (DIM)
- Modified damping impedance method (MDIM)

The equivalent circuit diagrams of the interface algorithms are shown in Figure 5.8.

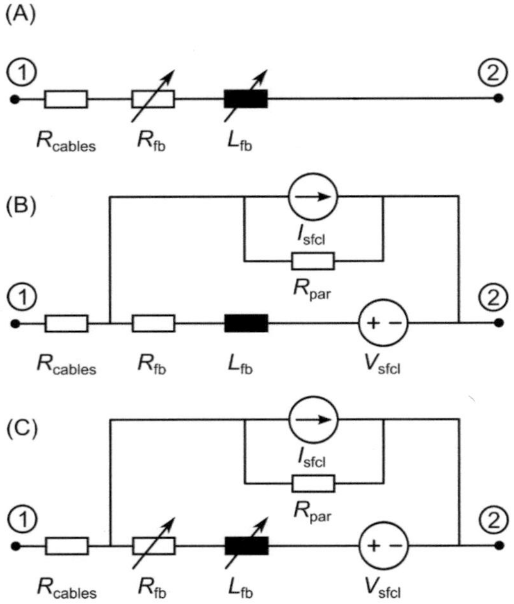

Figure 5.8: Equivalent circuit diagrams of (A) impedance feedback method, (B) classical damping impedance method and (C) modified damping impedance feedback method

The impedance feedback method represents the AC-SFCL only by a variable resistance R_{fb} and reactance X_{fb}. The values for resistance and inductance are instantaneous values calculated from the measured voltage and current using discrete Fourier transformation (DFT).

The damping interface method represents the AC-SFCL by a constant value for inductance and resistance adding a voltage source in series and a current source in parallel. The additional resistance R_{par} was used to suppress instabilities and was two orders of magnitudes higher than the impedance of the AC-SFCL in fault conditions. The resistance and inductance were set to the respective values of the AC-SFCL in normal condition. The voltage V_{sfcl} and the current I_{sfcl} are the measured instantaneous values of the voltage drop at the AC-SFCL demonstrator and the current conducted in the AC-SFCL demonstrator respectively.

The modified damping impedance algorithm combines the IFM and the DIM. In this case the AC-SFCL is represented by a variable resistance and inductance in series with a voltage source and a current source in parallel. The MDIM is equivalent to the DIM, except that the resistance and inductance are variable as in the IFM and calculated from instantaneous values of voltage V and current I.

For all three interface algorithms it is necessary to calculate the instantaneous values for resistance and inductance or reactance respectively. This was achieved by using discrete Fourier transformation (DFT) to extract the phase and magnitude of fundamental frequency components of the measured voltage and the measured current. This approach is shown in Figure 5.9.

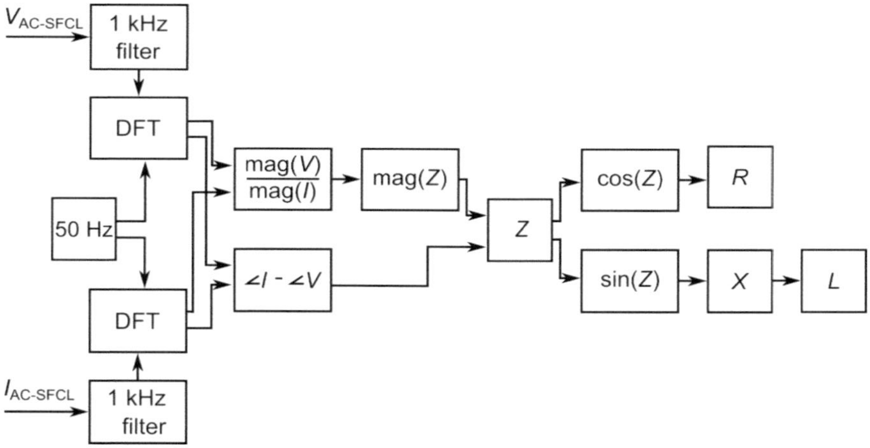

Figure 5.9: Approach to calculate instantaneous values of resistance and inductance from measured voltage and current

The measured voltage V_{sfcl} and current I_{sfcl} are passed through a 1 kHz filter to reduce noise. From the measured signals magnitude and phase are extracted using DFT and a 50 Hz reference signal. With the magnitude of the voltage V_{sfcl} and the current I_{sfcl} the magnitude of the impedance $|Z|$ is calculated. The phase difference between the voltage V_{sfcl} and the current I_{sfcl} is then used to calculate the complex impedance Z. Using trigonometry functions the resistance R and the reactance X and the inductance L are calculated from the impedance Z.

To compare the capability in terms of peak tracking all three interface algorithms have been implemented in the test setup shown in Figure 5.4 in order to reproduce the results for current limitation measured at KIT. For each interface algorithm the PHIL test setup was first fully simulated using the computational model of the AC-SFCL and then the actual PHIL measurement with the physical AC-SFCL connected to the AC-VVS was performed.

Figure 5.10 shows the limitation of a fault current of four half-cycles proceeded by a load cycle for the impedance feedback method. The prospective current was set to I_p = 2.0 kA. The reference measurement limits the current to 1.6 kA in the first half cycle and 1.48 kA in the following fault cycles. Using the IFM the simulated limited fault current in the first cycle shows numerical oscillation in the peak region < 200 A, preventing determination of a clear limited peak current. In the second cycle the limited peak is 30 A or 4% above the reference measurement with numerical oscillations < 40 A. In the third and fourth cycle the simulated current and the reference measurement deviate within < 2%.

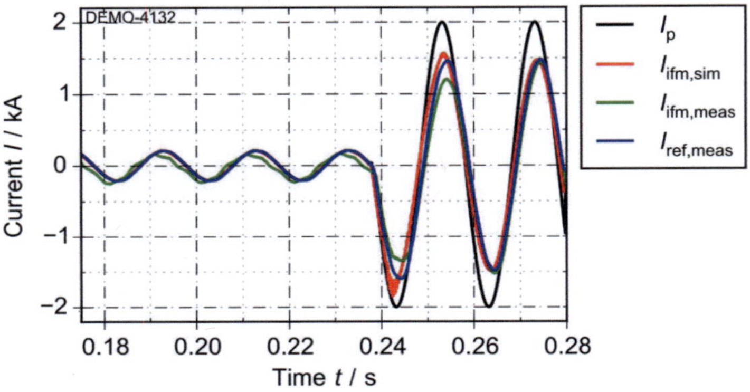

Figure 5.10: Comparison of results of the impedance feedback method (IFM) for simulation and measurement with the reference measurement during load and fault cycles

In the PHIL test of the IFM the fault current was limited to 1.35 kA in the first half cycle, which is significantly lower (10%) than the reference measurement. In the second half cycle the fault current is limited to 1.2 kA, which is 19% lower compared to the reference measurement. In the third and fourth half cycle the limited peak current of the PHIL measurement is within 3% of the reference measurement.

Figure 5.11 shows the limitation of a fault current of four half-cycles proceeded by a load cycle for the classical damping impedance feedback method. Using this interface algorithm no numerical oscillations occurred in the peaks of the limited currents in the simulated PHIL test setup. In the first half cycle the simulated limited fault current matches the reference measurement. In the following three half cycles the simulated limited fault current is within 4% of the reference measurement. In case of the PHIL measurement the limited peak in the first cycle is 1.46 kA, which is 9% lower than the reference measurement. This is the case in the third half cycle as well. In the second and fourth half cycle the measured limited peak current is within 4% of the reference measurement.

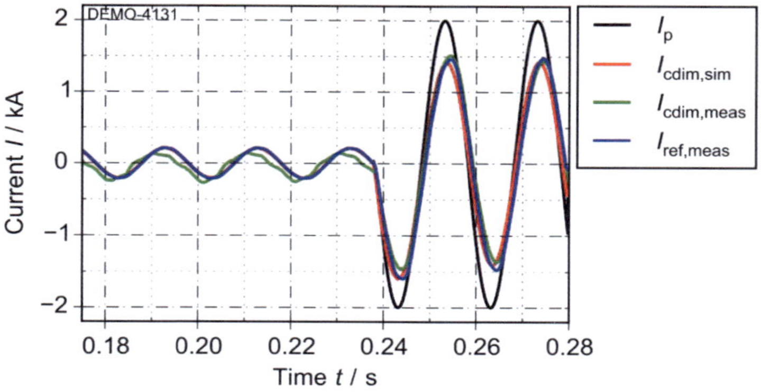

Figure 5.11: Comparison of results of the damping impedance feedback method for simulation and measurement with the reference measurement during load and fault cycles

Figure 5.12 shows the limitation of a fault current of four half-cycles proceeded by a load cycle for the modified damping impedance feedback method. In case of the simulated PHIL setup the limited peak currents deviate < 3% of the limited peaks of the reference measurement for all four half cycles. This constant deviation during the fault cycle was a well observed in the PHIL measurement, where the limited peak is within < 4% of the reference measurement.

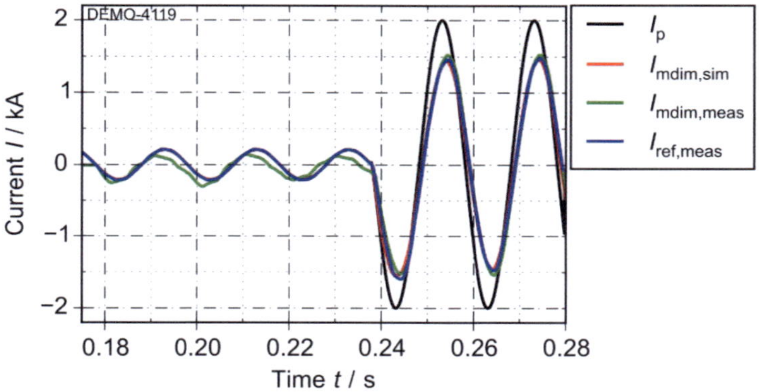

Figure 5.12: Comparison of results of the modified damping impedance feedback method for simulation and measurement with the reference measurement during load and fault cycles

The simulation and the measurement of the fault current limitation of the AC-SFCL demonstrator in the PHIL test setup has been performed for different prospective currents in the range of 0.5 kA to 2.5 kA. The results for the simulated PHIL test setup are summarized in Figure 5.13.

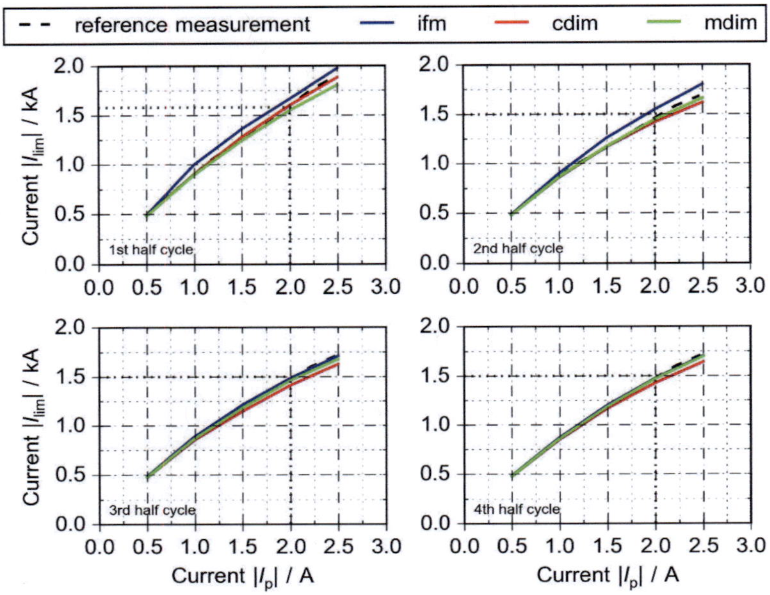

Figure 5.13: Comparison of current limitation of the three interface algorithms - simulation

In the first half cycle the DIM exhibits the best performance in terms of peak current tracking followed by the MDIM in peak fault current range. The IFM generally shows higher limited peak currents in the first half cycle as well as in the second half cycle. The DIM and the MDIM are matching the reference measurement in the second half cycle very well, with deviations < 4% at the upper and of the fault current range. In the third and fourth half cycle all interface methods show results which are within 4% of the reference measurement.

Figure 5.14 summarizes the results for PHIL measurement. In the first half cycle all interface algorithms show results, which are below the reference measurement with the MDIM being the closest to the reference measurement and the IFM deviating from the reference measurement by a margin of 10% throughout the whole peak fault current range. In the second half cycle the difference of reference measurement and PHIL measurement increases in the case of the IFM. Both, the CDIM and the MDIM, maintain values for limited fault current, which deviate by < 5% in the fault current range. In the third and the fourth half cycle the PHIL measurements of all interface algorithms are in good agreement with the reference measurement. The only exception is the CDIM, which exhibits deviations up to 15% in the third half cycle.

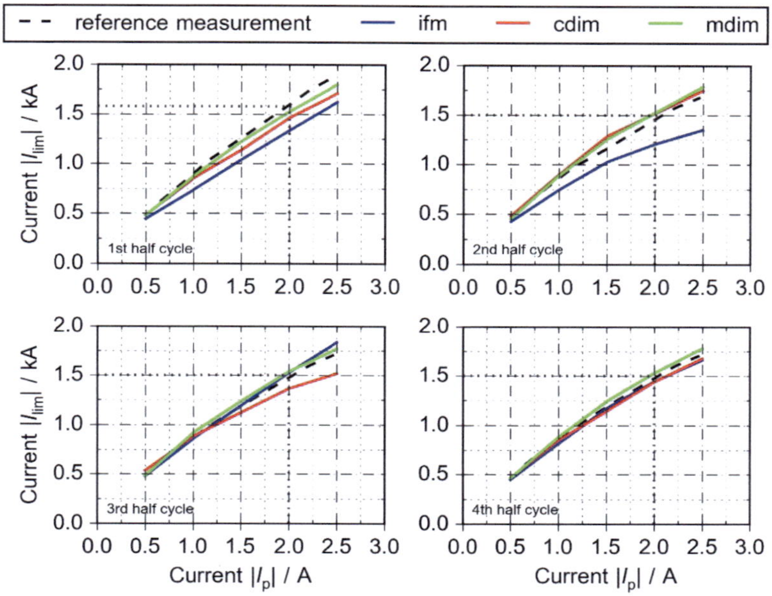

Figure 5.14: Comparison of current limitation of the three interface algorithms - measurement

Comparing the results for PHIL simulation and PHIL measurement all investigated interface algorithms exhibit different behavior in terms of peak tracking.

With deviations up to 15% in PHIL measurement compared to the reference measurement the impedance feedback algorithm shows the weakest peak tracking capability, especially during the important first peak. It was therefore discarded as interface algorithm for the verification of the PHIL setup and the 3-phase tests.

The classic damping impedance method generally shows a good agreement with the reference measurement in both PHIL simulation and PHIL measurement. However a significant deviation was observed in the third half cycle of the limited peak current during the fault.

Of all three interface algorithms the modified damping impedance method showed the most constant performance in terms of peak tracking in the PHIL simulation and the PHIL measurement. Therefore the MDIM was chosen as interface algorithm in the three phase tests.

5.3 3-phase short-circuit limitation

The tests and experiments described in chapter 5.1 and chapter 5.2 verified a stable and reliable operation of the AC-SFCL demonstrator in the PHIL test bed. The AC-SFCL demonstrator was originally designed as a single phase device. However existing commercial power systems are three phase systems.

Three phase systems can be subject to symmetrical and unsymmetrical short-circuits:

- Single line to ground
- Line to line fault with ground connection
- Three phase fault with ground connection
- Line to line fault without ground connection
- Three phase fault without ground connection

Using PHIL it was possible to simulate a three phase grid. Hereby the physical AC-SFCL demonstrator was inserted into phase A of the system. Phase B and phase C contained the respective interphase algorithm. Figure 5.15 shows the complete three-phase PHIL setup.

The power hardware part of the PHIL setup contains the AC-VVS and the 4.16 kV / 480 V three phase transformer T5. The AC-SFCL demonstrator was connected between the terminals A and B of T5. The current conducted in the demonstrator I_{SFCL} was measured as well as the voltage drop across the demonstrator V_{SFCL} and fed into the simulated system. The AC-SFCL demonstrator was fully submerged in liquid nitrogen under ambient pressure. The voltage drop was measured outside the liquid nitrogen bath resulting in an offset of 2 mΩ in the calculated resistance R_{sfcl}. The modified damping impedance method (chapter 5.2) was deployed as interface algorithm. Generally the configuration of the power hardware part remained the same as in the verification tests (chapter 5.1).

In real-time simulation environment a three phase system was set up. The system contains a fully transient model of a synchronous generator including voltage regulator and exciter. It was scaled down to a comparable power level of the AC-SFCL demonstrator. The parameters of the generator model are summarized in Table 5.1.

Each phase is connected to a load, which was used to adjust the rated current of the AC-SFCL in order to allow load cycles before or after the fault. Each phase is connected to a short-circuit path containing a variable resistor ($R_{flt\text{-}A}$, $R_{flt\text{-}B}$ and $R_{flt\text{-}C}$) to adjust the peak fault current. The short-circuit paths are interconnected, so that different kinds of faults can be applied. A variable resistor $R_{flt\text{-}g}$ is connected in series allowed faults with ground connection. The resistances R_{gg} and R_{lg} were inserted at the neutral points of the load and generator to allow short-circuits with ground connection (R_{gg} = R_{lg} = 1 mΩ) and without ground connection (R_{gg} = R_{lg} = 1 MΩ).

Table 5.1: Parameters of the generator model

Parameter	Value
Apparent power S_r	120 kVA
Rated Voltage V_r (L-L)	693 V
Rated Current I_r	100 A
Frequency f_n	50 Hz
X_a (stator leakage reactance):	0.130 pu
X_d (D-axis unsaturated reactance)	1.79 pu
X_d' (D-axis unsaturated transient reactance):	0.169 pu
X_d'' (D-axis unsaturated sub-transient reactance)	0.135 pu
X_q (Q-axis unsaturated reactance)	1.71 pu
X_q' (Q-axis unsaturated transient reactance)	0.228 pu
X_q'' (Q-axis unsaturated sub-transient reactance)	0.2 pu
R_a (stator resistance))	0.002 pu
T_{do}' (D-axis unsaturated transient open circuit time constant)	4.3 s
T_{do}'' (D-axis unsaturated sub-transient open circuit time constant)	0.032 s
T_{qo}' (Q-axis unsaturated transient open circuit time constant	0.85 s
T_{qo}'' (Q-axis unsaturated sub-transient open circuit time constant)	0.05 s

The AC-SFCL demonstrator is inserted in phase A of the system in series with the respective load. In order to replicate the behavior of the AC-SFCL in the phases B and C a variable impedance (resistance and inductance) is inserted.in series in each branch. The resistance and inductance are initialized with the respective values for normal operation of the AC-SFCL. If a phase is not subject to a fault during a test, the respective resistance and inductance will remain at values for normal operation. If a fault is triggered, which involves the phase or the branch, the resistance and inductance are updated based on the measurement of the actual AC-SFCL. With this approach it is possible to trigger faults for single phase, phase to phase and all three phases with and without ground connection.

During the tests several parameters were altered to investigate their influence. For the short-circuit test shown in Figure 5.16 to Figure 5.21, the total resistance was set to R_{flt} = 40 mΩ, the apparent power to S_{base} = 120 kVA , the power factor of the load set to p_f = 1 and the load was adjusted to enable the rated current for normal operation of the AC-SFCL. With this set of parameters, five types of faults were investigated. Before the fault was triggered for five cycles, three load cycles were triggered.

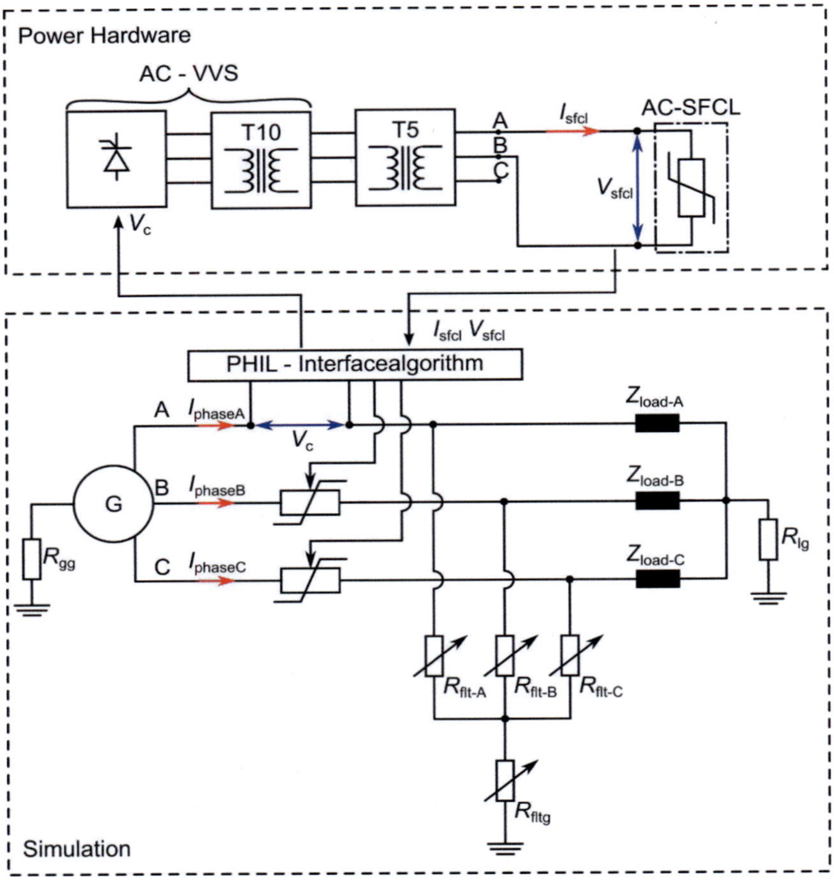

Figure 5.15: Three phase PHIL test setup. The Hardware part remained the same as in the preliminary tests shown in Figure 5.4

Figure 5.16 shows the results for a one phase to ground fault. The prospective current in the first cycle is limited from $|I_p|= 1.85$ kA to $\hat{I}_{sfcl} = 1.42$ kA (23.5%). In the following negative cycles the current limitation is decreasing with decreasing peaks and a last negative peak of 13.1% before the fault is switched-off. The prospective current and the limited current both show a decaying DC component, which is expected from a short-circuit next to a generator [Schw06], [Schl05], [Pis09]. As expected the currents of phases A and B remain unchanged. When the fault is turned off higher frequency oscillations (around 500 Hz) on the measured current I_{sfclA} are visible. These oscillations can be observed in the following measurements as well. They exclusively occur after switching off the fault and not within the measured load cycles and fault cycles. Therefore they have not been subject to further investigation within the scope of this work.

The voltage at the AC-SFCL is slightly phase shifted indicating an inductive current limitation. This becomes more evident looking at the instantaneous calculated values for resistance and reactance. After the fault starts the calculation needs a cycle to settle and deliver stable values. The resistance hereby reaches its steady-state design value of R_{sfcl} = 74 mΩ. The reactance remains well below its steady-state design value at X_{sfcl} = 75 mΩ. This indicates, that the induced current in secondary winding is still capable of partially shielding the primary winding and the *RE*BCO tapes are in a thermal stable state due to the copper stabilization.

Figure 5.17 shows the results for an unsymmetrical phase to phase fault with ground connection. The fault is set between phase A and phase B and the fault current limitation in this case is $|I_p|$= 1.79 kA limited to $|I_{sfcl}|$ = 1.37 kA. This is the same current limitation of 23.3% in the first half cycle as in the line to ground fault, which as well decreases with decreasing peaks. The simulated current in phase B increases as expected with triggering of the fault and reaches a peak of $|\hat{I}_{phaseB}|$ = 1.374 kA in the first half cycle and decreases with advancing fault as well, while the simulated current in the unaffected phase C remains at load level. The voltage is phase shifted and shows a decreasing peak in simulation and measurement during the fault, which leads to a decreasing impedance. While the resistance reaches the design value and is maintained, the reactance remains below the resistance and decreases until the end of the fault. The reason for this behavior is, that the generator is set to an apparent power of S_{base} =120 kVA and is therefore not capable of delivering enough power to keep the AC-SFCL demonstrator further in fault limiting operation. The decrease of resistance and reactance in Figure 5.17 (C) indicates a recovery of the *RE*BCO tapes during the fault.

Figure 5.18 shows the results for a symmetric three phase fault with ground connection. The current in all three phases increases as expected after triggering the fault peaking at $|I_{sfcl}|$ = 1.32 kA, $|I_{phaseB}|$ = 1.35 kA and $|I_{phaseC}|$ = 1.16 kA, while the prospective current reaches $|I_p|$ = 1.76 kA. For this kind of fault the resistance and reactance behave as in the 2-phase fault with ground connection. The resistance reaches its design value and decreases after a few cycles. The reactance stays below its design value and decreases after a few cycles to the same degree as the resistance, indicating a recovery of the *RE*BCO tapes.

The unsymmetrical phase to phase fault and the symmetrical three phase fault have been repeated without ground connection ($R_{flt\text{-}g}$ = 1mΩ). The results of these tests are shown in Figure 5.19 (phase to phase) and Figure 5.20 (3-phase). The results of the three phase faults without ground connection are similar compared to the respective faults with ground connection.

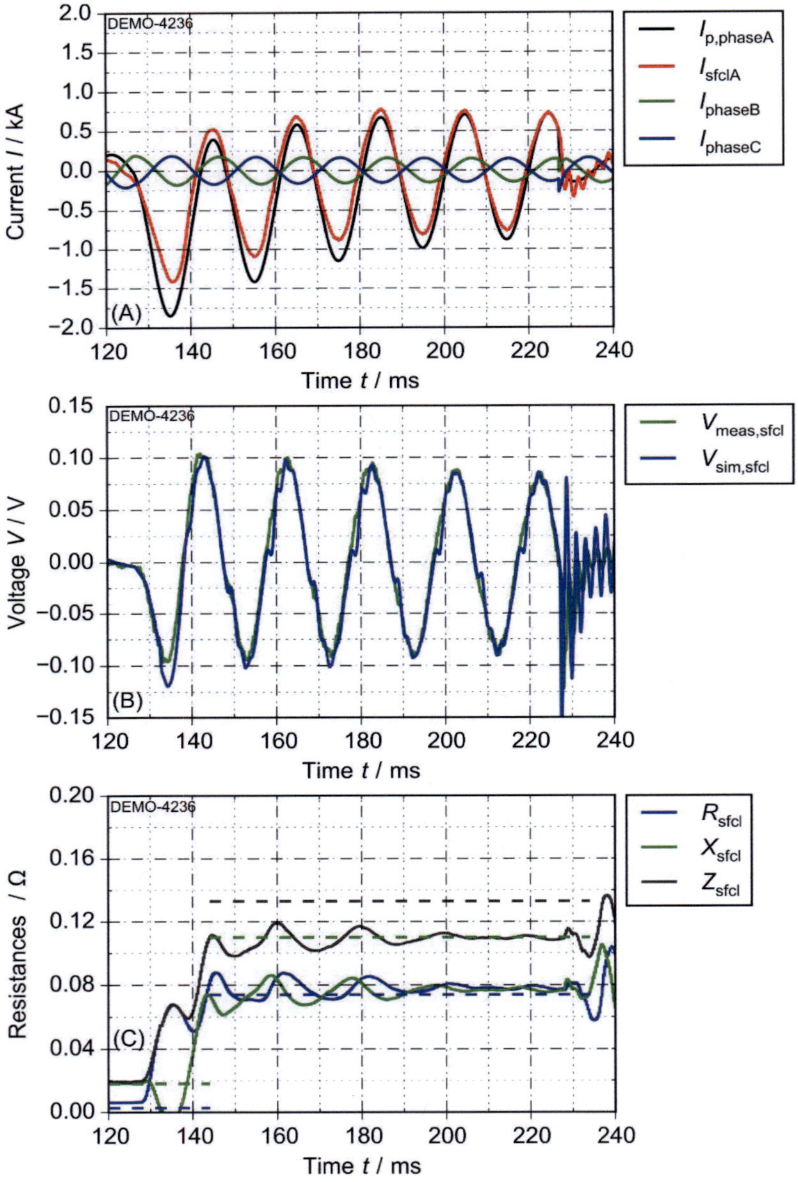

Figure 5.16: Results for fault current limitation of a 1-phase fault for (A) currents, (B) voltages and (C) resistances (S_{base} = 120 kVA, R_{flt} = 40 mΩ, p_f = 1.0)

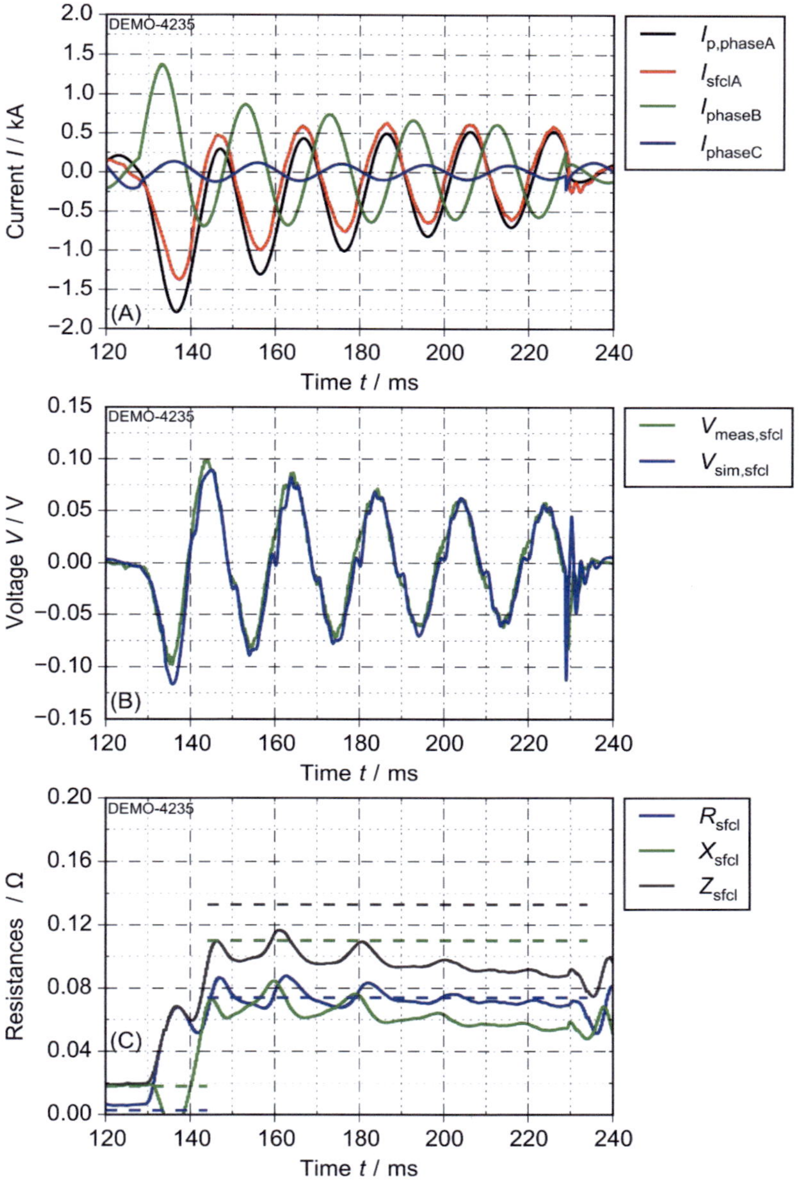

Figure 5.17: Results for fault current limitation of a 2-phase fault with ground connection for (A) currents, (B) voltages and (C) resistances (S_{base} = 120 kVA, R_{flt} = 40 mΩ, p_f = 1.0)

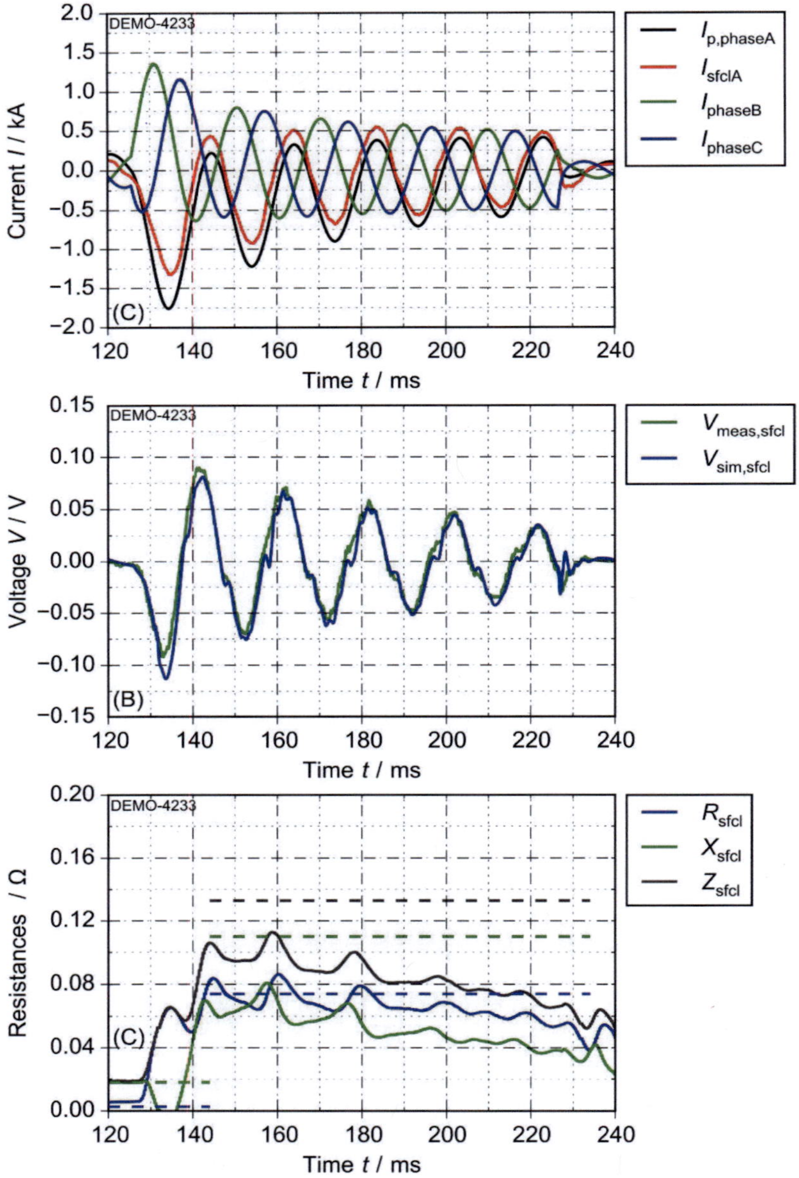

Figure 5.18: Results for fault current limitation of a 3-phase fault with ground connection for (A) currents, (B) voltages and (C) resistances (S_{base} = 120 kVA, R_{flt} = 40 mΩ, p_f = 1.0)

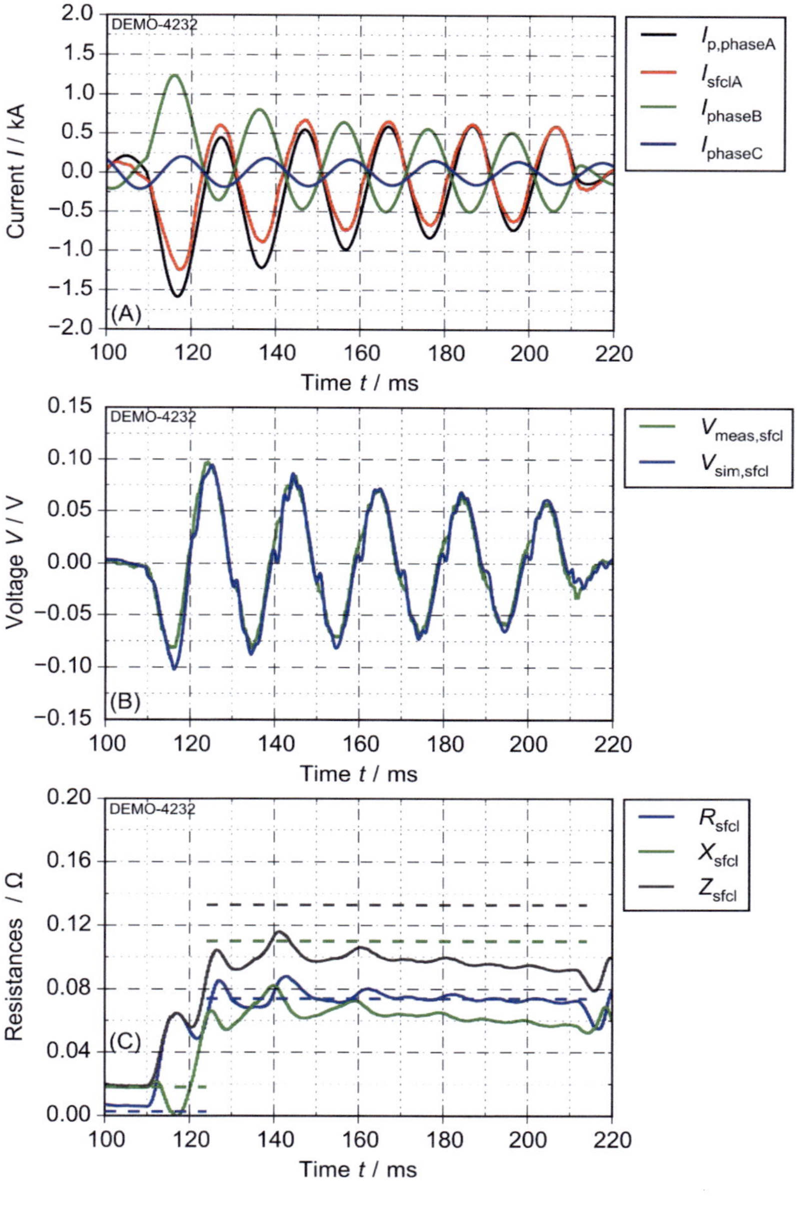

Figure 5.19: Results for fault current limitation of a 2-phase fault without ground connection for (A) currents, (B) voltages and (C) resistances (S_{base} = 120 kVA, R_{flt} = 40 mΩ, p_f = 1.0)

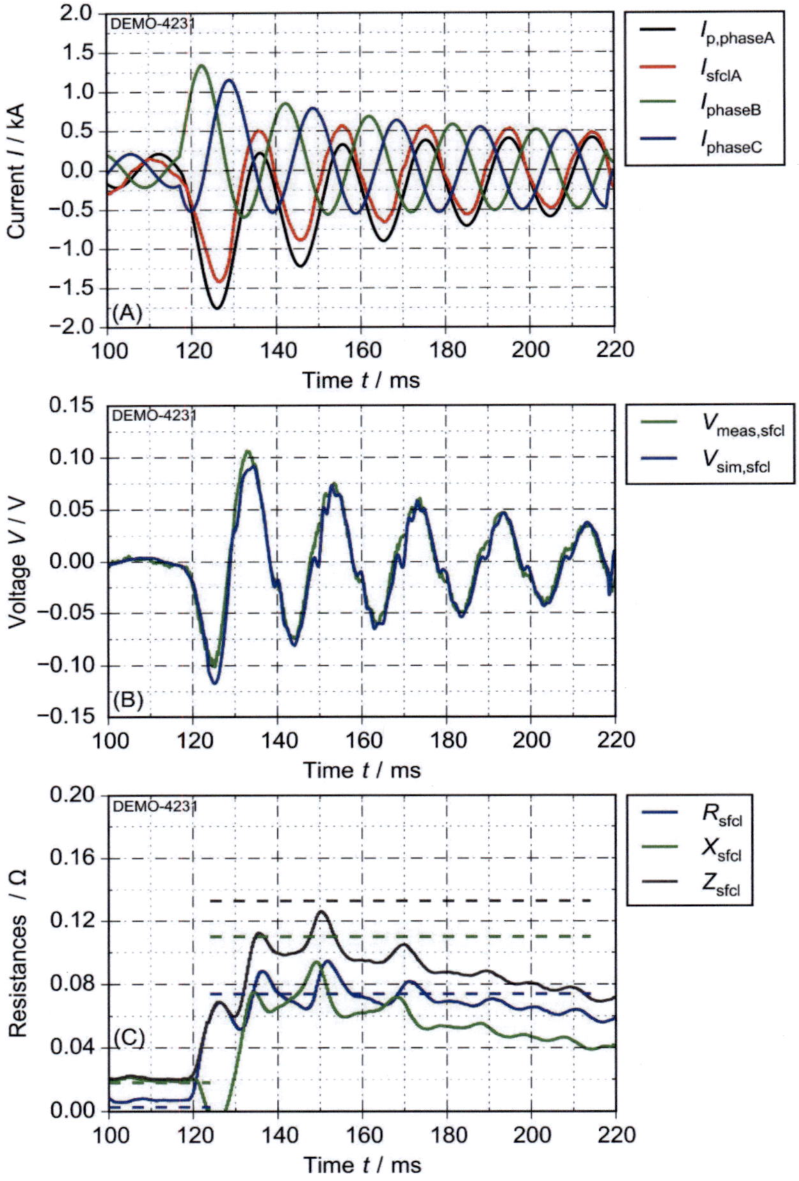

Figure 5.20: Results for fault current limitation of a 3-phase fault without ground connection for (A) currents, (B) voltages and (C) resistances (S_{base} = 120 kVA, R_{flt} = 40 mΩ, p_f = 1.0)

Variation of generator power S_{base}

Since the calculated resistance and reactance showed, that the AC-SFCL demonstrator started recovering during the fault cycle due to the limited power of the generator, the apparent power was increased from S_{base} = 120 kVA to S_{base} = 180 kVA and the fault duration increased to eight cycles to allow the instantaneous values for resistance and reactance to settle. The total fault resistance was set to R_{flt} = 40 mΩ, the power factor to p_f = 1 and the fault was preceded by three load cycles.

Figure 5.21 shows the results for the phase to phase fault. The prospective current in phase A is limited from $|I_p|$ = 2.23 kA to $|I_{sfcl}|$ = 1.78 kA (20%) in the first half cycle. In the third half cycle the current limitation is increasing to 28.1% (from $|I_p|$ = 1.63 kA to $|I_{sfcl}|$ = 1.17 kA) and then decreasing with decreasing peaks. The current in phase B reaches a peak of $|I_{phaseB}|$ =1.64 kA in the first half cycle and descending below 1 kA in the third half cycle. Like expected phase C is not affected by the fault and the current remains at load level. The measured and the simulated voltage at the AC-SFCL are in good agreement and show a phase shift and decreasing peaks. The calculated instantaneous value for the resistance settles approximately 50 ms after triggering the fault at R_{sfc} = 81 mΩ and maintains this values for the complete duration of the fault. The calculated reactance stays below the design value and decreases, after the calculation is settled, slightly from X_{sfcl} = 100 mΩ to X_{sfcl} = 92 mΩ. As a result the impedance decreases by the same degree from Z_{sfcl} = 130 mΩ to Z_{sfcl} = 122 mΩ.

The three phase fault with an apparent power of S_{base} = 180 kVA is shown in Figure 5.22. All phases are affected by the fault current and the currents increase as expected. The prospective current in phase A is limited from $|I_p|$ = 2.51 kA to $|I_{sfcl}|$ = 1.97 kA (21.6%) in the first half cycle. In the third half cycle the current limitation is increasing to 30.6% (from $|I_p|$ = 1.64 kA to $|I_{sfcl}|$ = 1.14 kA) and then decreasing with decreasing peaks. The currents in phase B and phase C reach peaks of $|I_{phaseB}|$ =1.85 kA and $|I_{phaseC}|$ =1.58 kA. In both simulated phases the current peaks decrease and match the measured current peaks in the physical phase A. The voltage drop at the AC-SFCL demonstrator shows a smaller phase shift as in the phase to phase fault and a stronger decrease in peaks.

Accordingly the impedance decreases during the fault (from Z_{sfcl} = 120 mΩ to Z_{sfcl} = 104 mΩ). As in the unsymmetrical case this decrease is exclusively caused by the decreasing reactance, since the resistance remains constant at R_{sfcl} = 80 mΩ.

Figure 5.23 shows the major peak currents for asymmetrical short-circuit in the measured phase A and simulated phase B together with the results from the short-circuit tests performed at KIT (compare Figure 4.30).

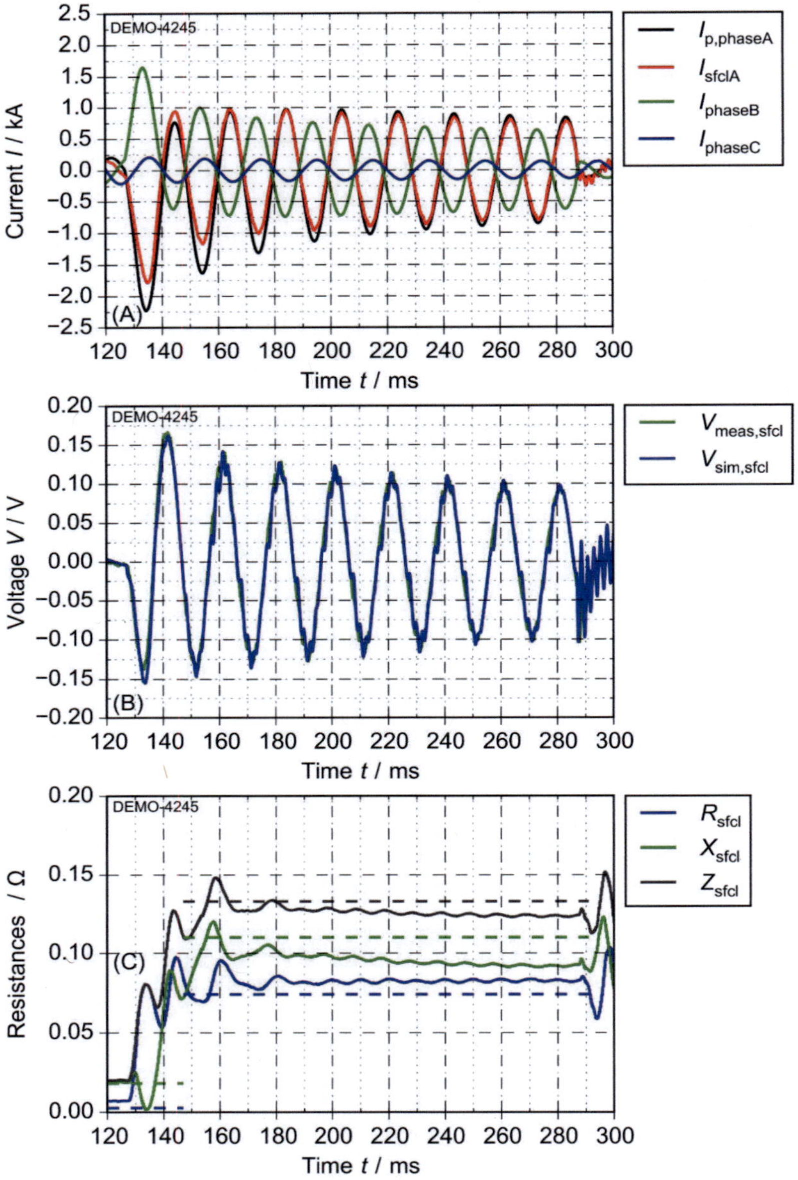

Figure 5.21: Results for fault current limitation of a 2-phase fault with ground connection for (a) currents, (b) voltages and (c) resistances (S_{base} = 180 kVA, R_{flt} = 40 mΩ, p_f = 1.0)

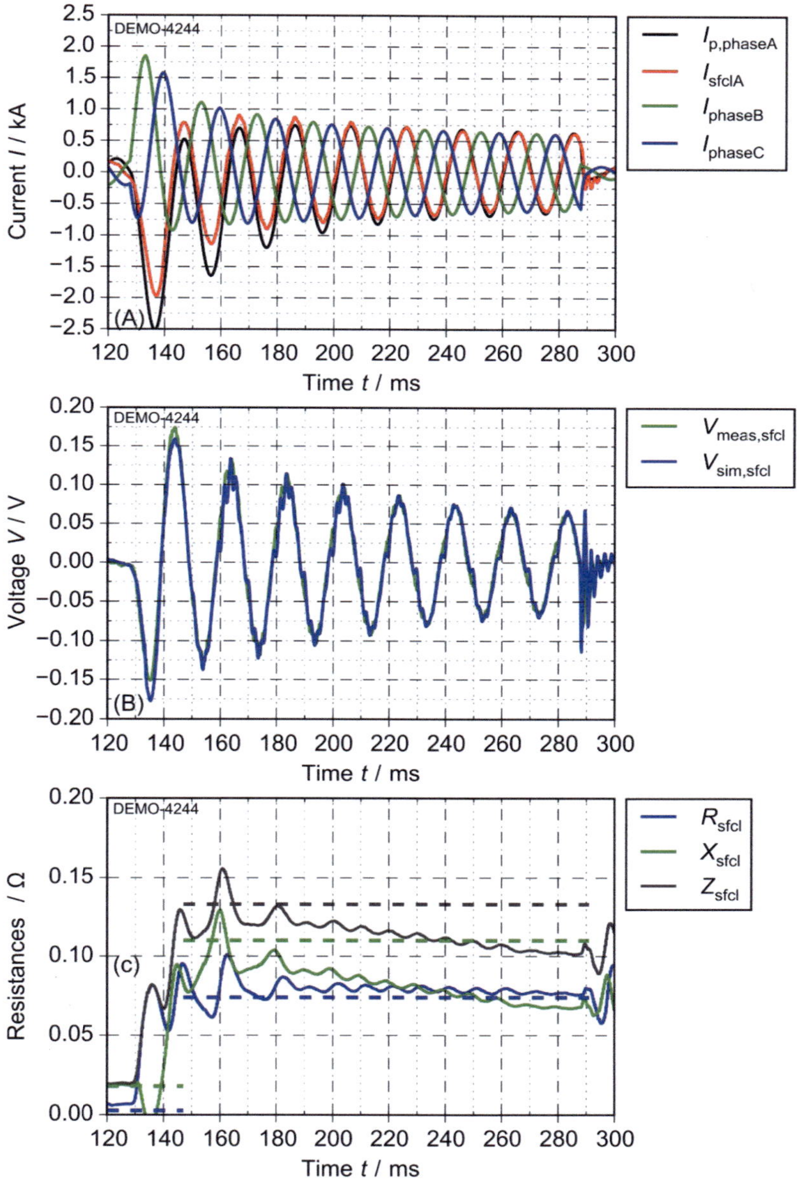

Figure 5.22: Results for fault current limitation of a 3-phase fault with ground connection for (a) currents, (b) voltages and (c) resistances (S_{base} = 180 kVA, R_{flt} = 40 mΩ, p_f = 1.0)

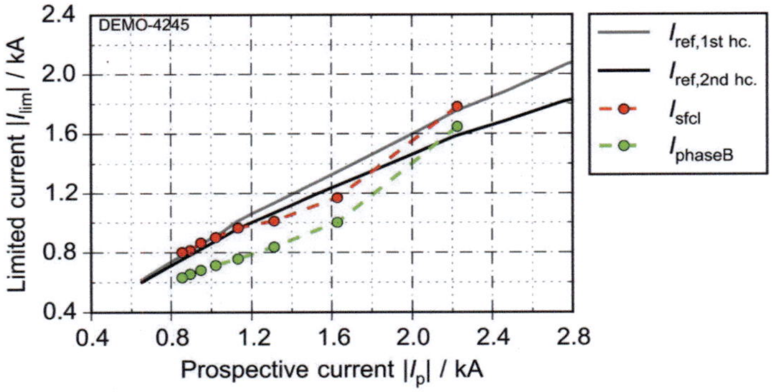

Figure 5.23: Summary of fault current limitation for unsymmetrical fault (S_{base} = 180 kVA)

The peaks of the limited fault current in the measured phase A are in very good agreement with the limited reference peak from single phase measurements. The measured current limitation is hereby in the first peak in good agreement with the first peak of the reference measurement and the following peaks in good agreement with the second peak of the reference measurement. The limited current peaks in the simulated phase B show the same progression as the measured peaks, but with a better limitation.

The major peaks for the symmetrical fault are shown in Figure 5.24 together with the results from the short-circuit tests performed at KIT (compare Figure 4.30). From the measurements it can be concluded, that the single phase AC-SFCL demonstrator shows the same fault current limitation capability in a three phase system as in single phase tests. The results obtained from the PHIL tests for fault current limitation are in good agreement with the single-phase short-circuit measurements for both symmetrical and asymmetrical short-circuits.

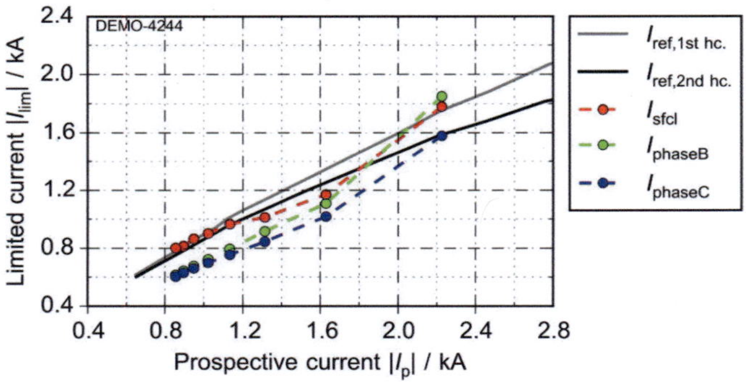

Figure 5.24: Summary of fault current limitation for symmetrical three phase fault (S_{base} = 180 kVA)

6 Conceptual Designs for Applications

The AC-SFCL demonstrator verified the principle of the AC-SFCL as well as the used design equations and design method respectively. Based upon these equations conceptual designs of the AC-SFCL for medium, high and ultra-high voltage are presented in this chapter. A single phase design is assumed for this work. For each voltage level a retrofit is calculated and compared to a calculated design with improved primary winding. Starting point is a conventional air core reactor, which is retrofitted with a secondary superconducting winding using the design approach described in chapter 3.4.1. This design process was implemented in the Mathcad software package for the calculation.

The impedance of the retrofit is depending on the maximum current I_{max}, the thickness of stabilization of the *RE*BCO tape h_{stab} and the distance between the windings d_w. All three parameters are varied in a reasonable value range. The objective is to find a specific design, which has a ratio of impedance during fault and impedance during normal operation of $z_{lim}/z_n = 5$ and an absolute impedance during normal operation of $z_n \leq 1\%$. This retrofit should have the maximum possible winding distance d_w and stabilization h_{stab} to match these conditions. The maximum current allowed in a *RE*BCO conductor is assumed to $I_{max} = 300$ A for all designs. Nevertheless the maximum current I_{max} is varied in order to show its influence on the impedance of the designs.

In a second step an improved design with lower impedance during normal operation is calculated. This improved design covers two cases:

- If the conditions are met by the retrofit, the low impedance design increases the ration of Z_{lim}/Z_n
- If the conditions are nor met by the retrofit, the low impedance design shows the geometry with the least volume meeting the conditions.

In order to calculate the low impedance design the inner radius r_{ip}, height h_p and number of turns N_p of the primary winding are varied to find all geometries which have the same inductance as the air core reactor in case of the retrofit using the design approach described in chapter 3.4.2 implemented in the Mathcad software package. The geometry with the least volume is chosen and a secondary superconducting winding is calculated.

Both designs are compared in terms of geometry, impedances, required conductor length and performance, losses and expected current limitation.

6.1 Medium voltage, 10 kV, 10 MVA, *z* = 6%

A voltage of 10 kV is a typical voltage level for distribution grids. Firstly a retrofit for a given air core reactor is designed by variation of the parameters of the secondary winding for 10 kV, 1 kA and z = 6%. Table 6.1 shows the main parameters of a typical air core reactor for medium voltage.

Table 6.1: Main parameters of the air core reactor for medium voltage [Scha13]

Parameter	Symbol	Value
Rated voltage	V_n	10 kV
Rated current	I_r	1 kA
Frequency	f_n	50 Hz
Reference impedance	z_{ref}	6%
Impedance	$\lvert Z_{ACR}\rvert$	0.3465 Ω
Reactance	X_{ACR}	0.3464 Ω
Inductance	L_p	1.103 mH
Resistance (primary winding)	R_p	7.2 mΩ
Inner diameter	d_{ip}	1104 mm
Outer diameter	d_{ap}	1184 mm
Height	h_p	990 mm
Number of turns	N_p	36.5
Conductor cross-section	A_{cop}	692 mm
Fill-factor	f_{fp}	0.6378
Current density	j_p	1.445 A / mm²

Retrofit of an air core reactor

In order to calculate the retrofit the distance of the windings d_w, the current I_{max} of the *RE*BCO tape and the thickness of the copper stabilizer h_{stab} have been varied. The current I_{max} is hereby the current, which every *RE*BCO tape must be capable of carrying independently of the position in the secondary winding and the applied magnetic field (compare eq. (3.91)). This approach allows to define the requirements in terms of in-field performance of the *RE*BCO tape. The current I_{max} was varied between 200 A and 600 A in steps in 50 A. The thickness of the copper stabilizer was varied from 0 µm to 100 µm in steps of 10 µm plus an additional silver layer of 2 µm for stabilization for all cases. The width of the *RE*BCO tape was constant for all the calculations at 12 mm. These values and value ranges are typical for current *RE*BCO tapes and cover performance improvements expected in the next years (compare chapter 2.1.3). The winding distance d_w was varied between 1 cm and 10 cm in steps of 1 cm. This range of d_w is assumed to be technically feasible to include the thermal and electrical insulation. The parameters and parameter ranges in order to calculate the retrofit are summarized in Table 6.2.

Figure 6.1 shows the calculated results for Impedance depending on winding distance d_w and stabilizer thickness h_{stab} at a constant maximum current of I_{max} = 300 A. The impedance Z_{lim} of the ACR is only matched, if the *RE*BCO conductor has no additional copper stabilization. With increasing copper stabilization the impedance during fault is decreasing due to the decreasing resistance R'_s (compare eq. (3.77)).

Table 6.2: Input parameters for calculation of the retrofit

Parameter	Symbol	Value (range)
Maximum current	I_{nax}	200 A - 600 A
Thickness of copper stabilizer	h_{stab}	0 µm – 100 µm
Thickness of silver stabilizer	h_s	2 µm
Winding distance	d_w	0 cm - 10 cm
Tape width	w_t	12 mm
Fill-factor secondary winding	f_{fs}	0.8

The impedance during normal operation is matching the z_n = 1% criterion at a winding distance of d_w =2 cm. This is the case independently of the amount of the copper stabilization, which has no effect during normal operation. In order to achieve the defined ration of z_{lim}/z_n = 5 the maximum copper stabilization is h_{stab} =10 µm.

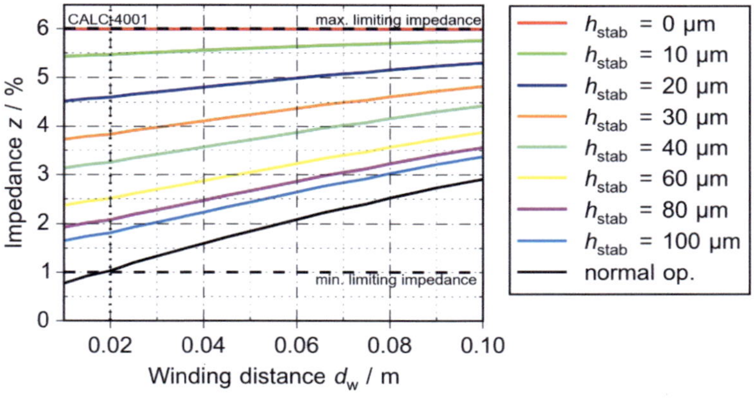

Figure 6.1: Calculated impedance of the retrofit for normal operation and fault condition for constant current I_{max} = 300 A

Figure 6.2 shows the calculated results for the impedance depending on the maximum current I_{max} and the stabilizer thickness h_{stab} for a constant winding distance of d_w = 2 cm. For constant stabilizer thickness h_{stab} the impedance during fault is increasing with increasing current with I_{max}. The increase of the impedance z_{lim} is expected, since with increasing current I_{max} the number of necessary *RE*BCO tapes to carry the induced current is decreasing and thus the total amount of copper in the secondary winding. Within the shown range the number of parallel *RE*BCO tapes n_{par} is more than sufficient to shield the primary winding completely.

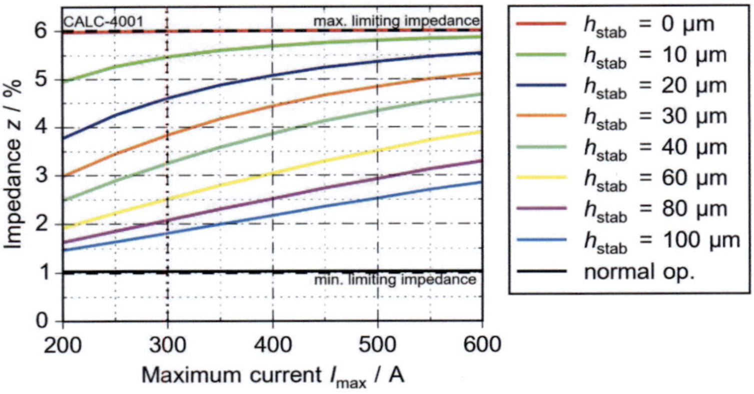

Figure 6.2: Calculated impedance of the retrofit for normal operation and fault condition for constant winding distance d_w = 2 cm

For constant stabilizer thickness h_{stab} the impedance during fault is increasing with maximum current I_{max} as well. This progression is expected for the impedance during fault z_{lim}, since the amount of *RE*BCO tapes is directly depending on the maximum current I_{max} (eq. (3.90)) and therefore the resistance R'_s (eq. (3.80)). The main electrical parameters relevant for operation of the retrofit are summarized in Table 6.3.

Table 6.3: Main electrical parameters of the calculated retrofit for 10 kV, 1 kA, z = 6%

Parameter	Symbol	Value	p.u.
Primary Resistance	R_p	7.2 mΩ	0.125%
Primary stray reactance	$X_{\sigma p}$	i58.9 mΩ	1.02%
Main reactance	X_m	i287.5 mΩ	4.98%
Secondary stray reactance	$X'_{\sigma p}$	i0.018 mΩ	0.31‰
Secondary resistance	R'_s	590 mΩ	7.4%
Impedance during fault	Z_{lim}	(140+257i) mΩ	5.07%
Impedance during normal operation	z_n	(7.2+58.9i) mΩ	1%

The defined impedance during load is reduced to z_n = 1%. During fault condition the retrofit has a lower impedance compared to the air core reactor, but the ratio of impedance during fault and during normal operation meets the defined condition of z_{lim}/z_n =5%.

Low impedance design

For the calculation of the primary winding with low impedance design the fill-factor f_{fp} and conductor cross-section A_{Cu} were assumed to be the same as in the case of the air core reactor. The secondary superconducting had the same distance of d_w = 2 cm from the primary winding. One layer of *RE*BCO tapes was assumed and the maximum current of each *RE*BCO tape was calculated accordingly. The additional copper stabilization was set to h_{stab} = 0 µm in order to maximize the resistance as a best case scenario for the impedance. This choice is purely theoretical as in real operation the *RE*BCO tape is very likely to burn out immediately. The input parameters for the calculation of the low impedance design are summarizes in Table 6.4.

Table 6.4: Input parameters for calculation of the low impedance design

Parameter	Symbol	Value (range)
Fill-factor primary winding	f_{fs}	0.6378
Conductor cross-section	A_{Cu}	692 mm
Winding distance	d_w	2 cm
Stabilization	h_{stab}	0 µm
Inner radius	r_{ip}	0.2 m – 0.8 m
Height	h_p	0.4 m – 1.6 m

Every calculated primary winding has the same impedance during fault as the air core reactor of Z_{ACR} = 0.364 Ω. Figure 6.3 shows the calculated results for impedance during normal operation depending on number of turns N_p, inner radius r_{ip} and height h_p of the primary winding

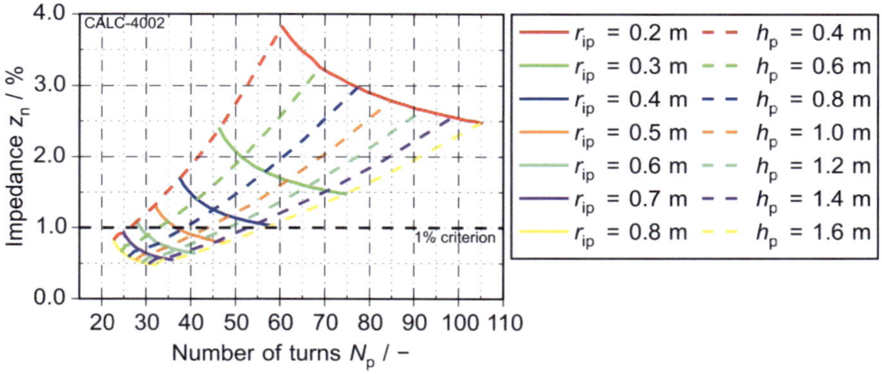

Figure 6.3: Calculated impedance during normal operation of 10 kV,1 kA, z = 6% AC-SFCLs for variation of the geometry of the primary winding

Figure 6.4 shows a magnified view of the calculated results for normal operation. In order to achieve a impedance during normal operation of $z_n \leq 1\%$ the inner radius must be higher than $r_i > 0.5$ m. Correspondingly a height of $h_p > 0.4$ m is required. Above those values certain combinations of height and inner radius allow geometries, which ensure a impedance during normal operation of $z_n < 1\%$.

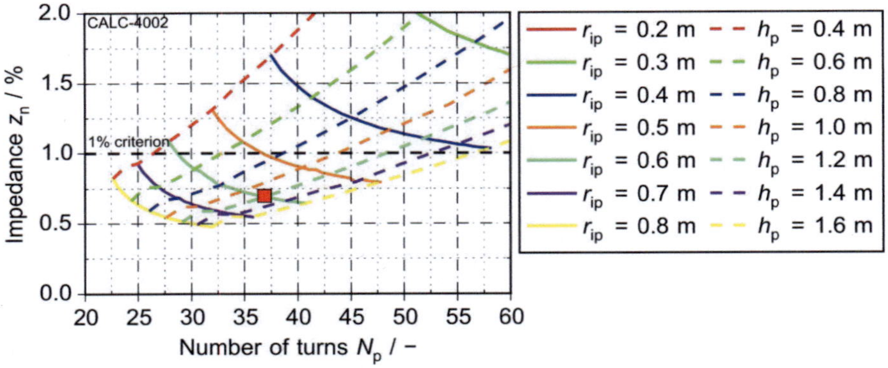

Figure 6.4: Magnified view of Figure 6.3. The red square indicates the chosen low impedance design

For the chosen geometry of the primary winding a secondary winding was calculated as in the case of the retrofit. Figure 6.5 shows the calculated results for Impedance depending on winding distance d_w and stabilizer thickness h_{stab} at a constant maximum current of I_{max} = 300 A.

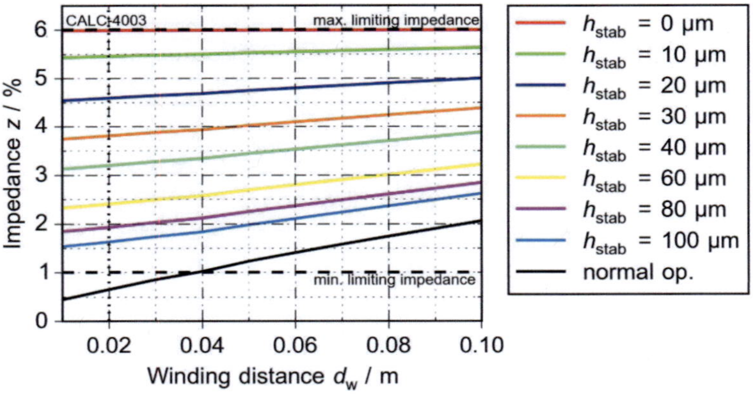

Figure 6.5: Dependency of impedance on winding distance and stabilization for the low impedance design (I_{max} = 300 A)

For the retrofit, the low impedance only matches the impedance of the air core reactor if no additional stabilization is applied on the *RE*BCO tapes. Assuming the same winding distance of d_w = 2 cm as in the case of the retrofit a stabilizer thickness of h_{stab} = 10 µm allows a ratio of impedance during fault and impedance during normal operation of z_{lim}/z_n = 6.

Figure 6.6 shows the dependency of the impedance during fault and normal operation of the design with low impedance on maximum current I_{max} and stabilizer thickness h_{stab}.

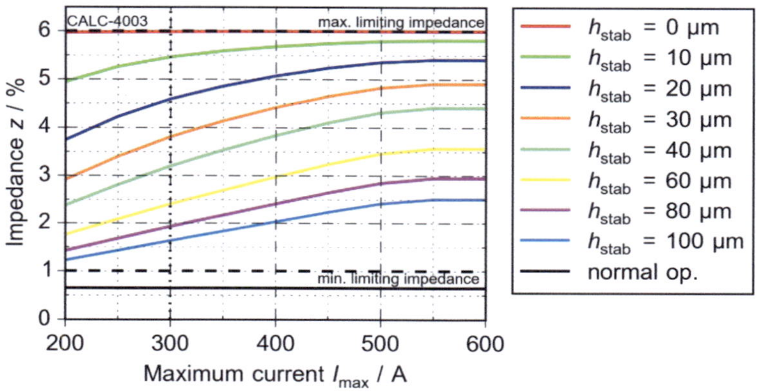

Figure 6.6: Dependency of impedance on critical current and stabilization for the low impedance design (d_w = 2 cm)

The impedance is increasing with maximum current until I_{max} = 550 A is reached. At this maximum current the number of needed *RE*BCO tapes allow shielding the primary winding with one layer of *RE*BCO tapes, which is a significant improvement compared to the retrofit.

Table 6.5 summarizes the main electrical parameters of the low impedance design. The resistance of the primary winding R_p remains almost unchanged. The stray reactances are lower compared to the retrofit, while the main inductance is increased due to the increased dimensions. Because of the increased height h_p more *RE*BCO tapes are needed in parallel to shield the primary winding, which lowers the resistance during fault.The AC-SFCL low impedance design reduces the impedance during load further than the retrofit from z_n = 1% to z_n = 0.6. At the same time the impedance during fault is reduced from z_{lim} = 5.2% to z_{lim}= 4.8%. However the ratio of impedance during fault to impedance during normal operation is increasing from z_{lim}/z_n = 5.2% to z_{lim}/z_n = 6. Besides the main electrical parameters several other parameters change, which are discussed in the next sub-chapter.

Table 6.5: Main electrical parameters of the low impedance design

Parameter	Symbol	Value	p.u.
Primary Resistance	R_p	7.76 mΩ	0.13%
Primary stray reactance	$X_{\sigma p}$	i37.4 mΩ	0.65%
Main reactance	X_m	i309 mΩ	5.35%
Secondary stray reactance	$X'_{\sigma s}$	i0.008 mΩ	0.00014%
Secondary resistance	R'_s	306 mΩ	9.76%
Impedance during fault	Z_{lim}	(126.5+290i) mΩ	5.47%
Impedance during normal operation	z_n	(7.8+37.4i) mΩ	0.66%

Comparison of calculated results

The main parameters of the retrofit and the low impedance design are summarized in Table 6.6.

Table 6.6: Comparison of retrofit and low impedance design

Parameter		Symbol	Retrofit	Low imp. design
Height		h_p	990 mm	1200 mm
Inner diameter		d_{ip}	1101 mm	1200 mm
Outer diameter		d_{ap}	1184 mm	1267 mm
Number of turns		N_p	36.5	36.9
Maximum field at	center	$B_{max\parallel}$	37 mT	31 mT
	end	$B_{max\perp}$	27 mT	25 mT
Number of tapes		n_{tot}	173	174
Maximum current		I_{max}	300 A	300 A
*RE*BCO length		l_{tot}	579 m	646 m
		l_{turn}	3.34 m	3.64 m
Stabilizer thickness		h_{stab}	10 µm	10 µm
Total AC-losses		P_{tot}	73.94 W	99.05 W
Impedance	normal op.	Z_n (z_n)	59.3 mΩ (1%)	38.1 mΩ (0.66%)
	fault op.	Z_{lim} (z_{lim})	293 mΩ (5.07%)	316 mΩ (5.47%)

The low impedance design possesses increased dimensions compared to the retrofit with slightly thinner wall thickness of the primary winding.

The increased dimensions results in increased amount of conductor material. The piece length is increased due to the increased diameter and more *RE*BCO tapes are needed to shield the primary winding due to the increased height. However the increased dimensions lead to lower magnetic field densities, which is beneficial for the requirements of the *RE*BCO tapes. The AC-losses of the low impedance design are higher compared to the AC losses of the retrofit by 25.09 W (25.4%), because of the increased tape length. The AC-losses of the retrofit and the low impedance design are summarized in Table 6.7. The major cause for AC losses are the losses due to self-field.

Table 6.7: AC-losses of the retrofit and the low impedance design

Parameter	Retrofit	Low impedance design
External perpendicular field	0.252 W	0.496 W
External parallel field	5.76 mW	6.55 mW
Self-field	73.89 W	98.52 W
Eddy current	0.015 W	0.028 W
Sum	73.94 W	99.05 W

In order to investigate the current limiting capability of the retrofit and the low impedance design calculations of steady-state short-circuit currents and limited short-circuit currents have been performed with the equations described in chapter 4.3. For the retrofit and the low impedance design the prospective current of I_p = 20 kA is limited to $|\hat{I}_{lim}|$ = 15.1 kA. This corresponds to a limitation of 24.5% in both cases. This means, that the slightly lower impedance during fault of the low impedance design of z_{lim} =4.8% compared to the impedance during fault of the retrofit of z_{lim} = 5.2% has a negligible influence on the fault current limitation. Figure 6.7 shows the limitation of fault currents for different prospective currents.

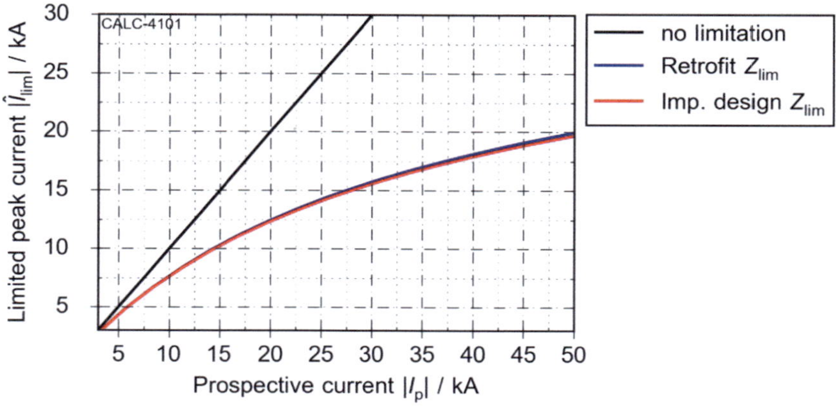

Figure 6.7: Expected fault current limitation of the retrofit and the low impedance design for 10 kV, 1 kA, z = 6%

The retrofit and the low impedance design show the same behavior in terms of current limitation: The current limitation increases with increasing prospective current. This behavior is similar to the current limitation of the ACSFCL demonstrator.

6.2 High voltage, 110 kV, 110 MVA, z = 6%

Voltages of 110 kV can be found in transmission networks. The approach to design an AC-SFCL is the same as in the case of medium voltage: Calculate the retrofit of an air core reactor, which has an impedance during normal operation of $z_n \leq 1\%$ and at the same time a significant increased impedance during fault operation. The main parameters of a typical air core reactor for 110 kV are summarized in Table 6.8.

Table 6.8: Main parameters of the air core reactor for high voltage [Scha13]

Parameter	Symbol	Value
Rated voltage	V_n	110 kV
Rated current	I_r	1 kA
Frequency	f_n	50 Hz
Reference impedance	z_{ref}	6%
Impedance	$\lvert Z_{ACR}\rvert$	3.8107 Ω
Reactance	X_{ACR}	3.8105 Ω
Inductance	L_p	12.129 mH
Resistance (primary winding)	R_p	35 mΩ
Inner diameter	d_{ip}	1854 mm
Outer diameter	d_{ap}	2160 mm
Height	h_p	1210 mm
Number of turns	N_p	84.75
Conductor cross-section	A_{cop}	597
Fill-factor	f_{fp}	0.2733
Current density	j_p	1.675 A / mm²

Retrofit of an air core reactor

In case of the retrofit for high voltage the parameters varied are the same as for medium voltage: The windings d_w, the current I_{max} of the *RE*BCO tape and the thickness of the copper stabilizer h_{stab}. Since the air core reactor for high voltage possess more turns as the air core reactor for medium voltage the induced current is assumed to be higher and thus the range of the maximum current I_{max} was extended. The other input parameters for the calculation remained the same or within the same value range respectively. The input parameters are summarized in Table 6.9.

The calculated results for impedance during normal operation and during fault operation depending on winding distance d_w and stabilizer thickness h_{stab} for a maximum current I_{max} = 300 A are shown in Figure 6.8. Even without additional copper stabilization the impedance of the air core reactor of z = 6% is not matched, but very close with 5.9%. Additional copper stabilization reduces the impedance significantly and above h_{stab} =40 µm the increase of impedance during fault is below 50%.

Table 6.9: Input parameters for calculation of the retrofit for high voltage

Parameter	Symbol	Value (range)
Maximum current	I_{nax}	200 A - 600 A
Thickness of copper stabilizer	h_{stab}	0 µm – 60 µm
Thickness of silver stabilizer	h_s	2 µm
Winding distance	d_w	1 cm - 10 cm
Tape width	w_t	12 mm
Fill-factor secondary winding	f_{fs}	0.8

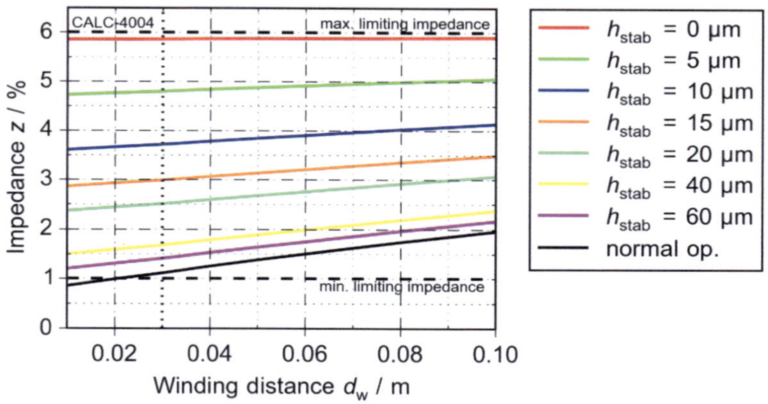

Figure 6.8: Calculated impedance of the retrofit for normal operation and fault condition for constant current I_{max} = 300 A

The impedance during fault is decreasing with increasing copper stabilization. With a winding distance d_w = 3 cm the retrofit almost fulfills the z = 1% criterion during normal operation. With a thickness of h_{stab} = 5 µm of additional copper stabilization the ratio of impedance during fault and impedance during normal operation is z_{lim}/z_n = 4.33%. These values are in principle acceptable for the retrofit, but do not meet the specifications.

Figure 6.9 shows the calculated results for impedance during normal operation and during fault operation depending on maximum current I_{max} and stabilizer thickness h_{stab} for a constant winding distance of d_w = 3 cm. The impedance during fault without copper stabilization matches the impedance of the air core reactor of z = 6% for maximum currents above 450 A. The impedance during fault is increasing for increasing maximum current for all stabilizer thicknesses h_{stab} > 0 µm in the shown range. This is caused by the decreasing demand of *RE*BCO tapes with increasing maximum current I_{max}. This number of *RE*BCO tapes is maintained with increasing maximum current I_{max} in order to ensure an optimal shielding.

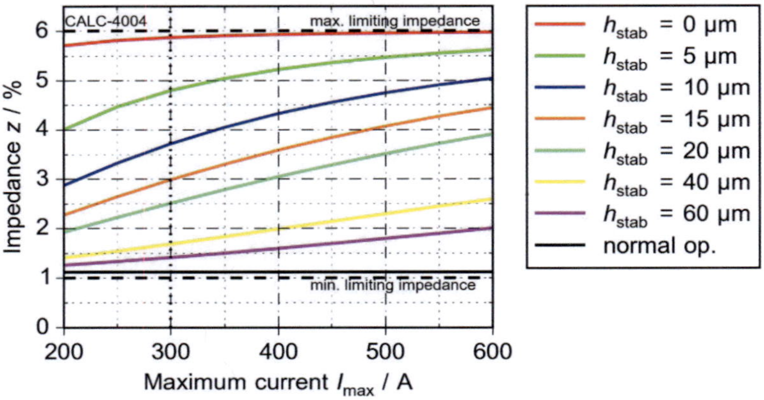

Figure 6.9: Calculated impedance of the retrofit for normal operation and fault condition for constant winding distance d_w = 3 cm

The main electrical parameters are for a retrofit with a maximum current I_{max} = 300 A, winding distance d_w = 3 cm and additional copper stabilization of h_{stab} = 5 µm are summarized in Table 6.10.

Table 6.10: Main electrical parameters of the calculated retrofit for 110 kV, 1 kA, z = 6%

Parameter	Symbol	Value	p.u.
Primary Resistance	R_p	35 mΩ	0.055%
Primary stray reactance	$X_{\sigma p}$	i0.707 Ω	1.11%
Main reactance	X_m	i3.1 Ω	4.89%
Secondary stray reactance	$X'_{\sigma p}$	i0.33 mΩ	0.5‰
Secondary resistance	R'_s	4.98 Ω	6.27%
Impedance during fault	Z_{lim}	(1.54+2.64) Ω	4.81%
Impedance during normal operation	z_n	(0.035+0.707) Ω	1.11%

Low impedance design

The input parameters for the low impedance design for high voltage are shown in Table 6.11 and are the same as in the case of medium voltage. The fill-factor of the primary winding f_{fp} and conductor cross-section in the primary winding A_{Cu} remain the same compared to the air core reactor. The winding distance is defined as d_w = 3 cm and the value ranges for inner radius r_{ip} and height h_p are adjusted to the increased dimension of the air core reactor. The number of turn in the primary winding N_p are calculated in such a way, that every geometry has the same impedance as the air core reactor (compare chapter 3.4.2).

Table 6.11: Input parameters for calculation of the low impedance design

Parameter	Symbol	Value (range)
Fill-factor primary winding	f_{fs}	0.2733
Conductor cross-section	A_{Cu}	597 mm
Winding distance	d_w	3 cm
Stabilization	h_{stab}	0 µm
Inner radius	r_{ip}	0.5 m – 1.0 m
Height	h_p	1.0 m – 2.0 m

The calculated results for impedance during normal operation depending on the geometrical parameters and the number of turns N_p are shown in Figure 6.10. In order to meet the criterion of $z_n \leq 1\%$ the inner radius must be at least $r_{ip} \geq 0.9$ m and the height $h_p \geq 1$ m at the same time.

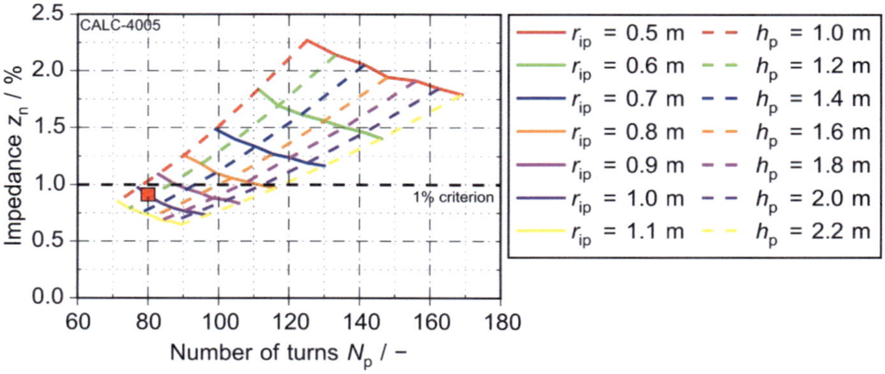

Figure 6.10: Results for impedance during normal operation of 110 kV,1 kA, z = 6% AC-SFCLs for variation of the geometry of the primary winding. The red square indicates the chosen low impedance design

For the low impedance design a rather modest increase of inner radius to r_{ip} = 1 m is sufficient to lower the impedance during normal in order to meet the $z_n \leq 1\%$ criterion. The height can remain the same as the retrofit at h_p = 1.2 m. For the primary winding with r_{ip} = 1 m and h_p = 1.2 m a secondary winding was calculated.

Figure 6.11 shows the results for impedance during normal operation and impedance during fault operation depending on winding distance d_w and stabilizer thickness h_{stab} at a constant maximum current of I_{max} = 300 A. The impedance during normal operation is reduced at a winding distance of d_w = 3 cm to z_n = 0.84%. Assuming the same amount of additional copper stabilization of h_{stab} = 5 µm the ratio of impedance during fault to impedance during normal operation is Z_{lim}/Z_n = 5.64.

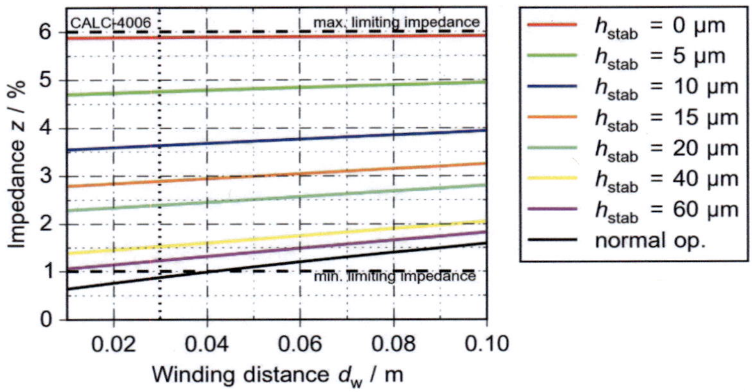

Figure 6.11: Dependency of the calculated impedance on winding distance and stabilization for the low impedance design (I_{max} =300 A)

Figure 6.12 show the dependency of the impedance during normal operation and during fault operation of the low impedance design on maximum current I_{max} and stabilizer thickness h_{stab} for a constant winding distance d_w = 3 cm. The impedance during fault operation z_{lim} is increasing with increasing maximum current. The reduction of the maximum current I_{max} with increasing dimensions corresponds with the results for medium voltage.

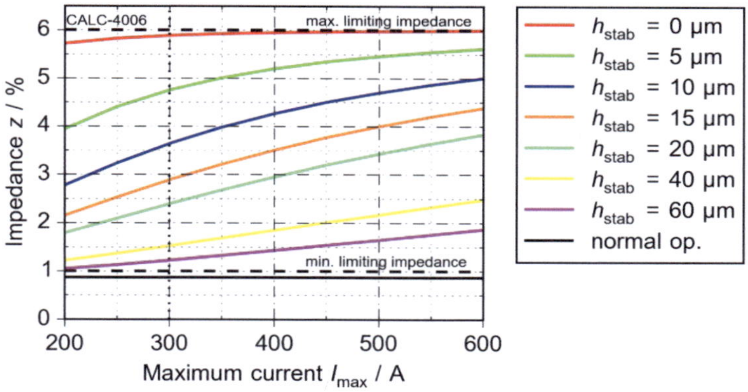

Figure 6.12: Dependency of the calculated impedance on critical current and stabilization for the low impedance design (d_w = 3 cm)

Table 6.12 summarizes the main electrical parameters of the low impedance design. The resistance of the primary winding remains unchanged and is negligible compared to the primary stray reactance $X_{\sigma p}$, which mainly determines the impedance during normal operation.

Table 6.12: Main electrical parameters of the low impedance design

Parameter	Symbol	Value	p.u.
Primary Resistance	R_p	38 mΩ	0.06%
Primary stray reactance	$X_{\sigma p}$	i0.535 Ω	0.84%
Main reactance	X_m	i3.28 Ω	5.17%
Secondary stray reactance	$X'_{\sigma s}$	i0.049 mΩ	0.07‰
Secondary resistance	R'_s	4.08 Ω	6.43%
Impedance during fault	Z_{lim}	(1.64+2.53i) Ω	4.74%
Impedance during normal operation	z_n	(0.038+0.535i) Ω	0.84%

The impedance during normal operation of the retrofit of z_n = 1.11% is reduced to z_n = 0.84%. At the same time the impedance during fault operation is negligible.

Comparison of results

The main parameters of the retrofit and the low impedance design are summarized in Table 6.13. The slight increase in diameter from d_{ip} = 1.854 m of the retrofit to d_{ip} = 2 m of the low impedance design reduces the maximum parallel magnetic field density. The increase in diameter also increases the required conductor length as well as the increased number of parallel *RE*BCO tapes. Both designs allow a copper stabilization of h_{stab} = 5 µm and nearly the same impedance during fault.

Table 6.13: Comparison of retrofit and low impedance design

Parameter		Symbol	Retrofit	Low imp. design
Height		h_p	1210 mm	1200 mm
Inner diameter		d_{ip}	1854 mm	2000 mm
Outer diameter		d_{ap}	2160 mm	2298 mm
Number of turns		N_p	84.75	81.875
Maximum field at	center	$B_{max\parallel}$	63 mT	59 mT
	end	$B_{max\perp}$	38 mT	38 mT
Number of tapes		n_{tot}	212	234
Maximum current		I_{max}	300 A	300 A
*RE*BCO length		l_{tot}	1196 m	1428 m
		l_{turn}	5.64 m	6.1 m
Stabilizer thickness		h_{stab}	5 µm	5 µm
Total AC-losses		P_{tot}	166.3 W	170.2 W
Impedance	normal op.	Z_n (z_n)	707 mΩ (1.11%)	535 mΩ (0.84%)
	fault op	Z_{lim} (z_{lim})	3.05 Ω (4.81%)	3.01 Ω (4.74%)

The low impedance design shows a higher ratio of impedance during fault and impedance during normal operation of Z_{lim}/Z_n = 5.62 compared to the retrofit with Z_{lim}/Z_n = 4.31. The AC-losses of the low impedance design are slightly higher by 2.36% (3.9 W) compared to the retrofit. As shown in Table 6.14 this is mainly determined by the self-field losses.

Table 6.14: AC-losses of the retrofit and the low impedance design

Parameter	Retrofit	Low impedance design
External perpendicular field	13.72 W	5.37 W
External parallel field	9.69 mW	0.01 W
Self-field	152.3 W	164.8 W
Eddy current	0.084 W	0.076 W
Sum	166.3 W	170.2 W

The steady-state fault current limitation was calculated for the retrofit and for the low impedance design using the equations described in chapter 4.3. Since the impedance during fault is nearly the same for both designs (2% difference), the fault current limitation is the same for both designs.

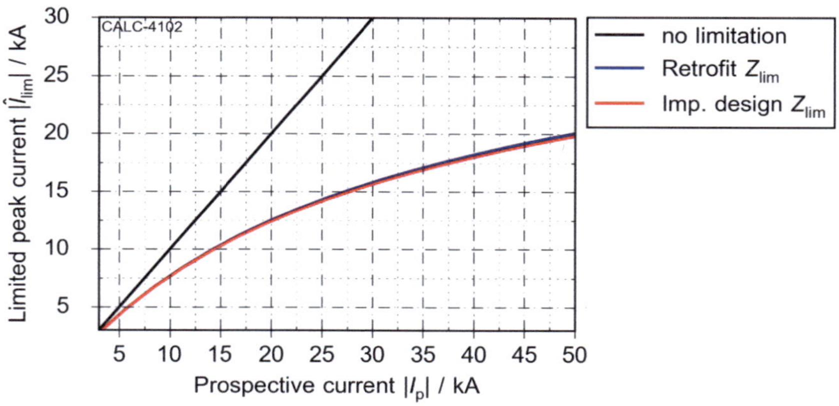

Figure 6.13: Expected fault current limitation of the retrofit and the low impedance design for 110 kV, 1 kA, z = 6%

Both designs limit a fault current to the same degree and the fault current limitation increases with the prospective current. A fault current of I = 20 kA ($20 \cdot I_n$) will be limited by 25% to I_{lim} = 15 kA for example.

6.3 Ultra-high voltage, 380 kV, 380 MVA, z = 6%

Voltages of V = 380 kV are used for long distances in transmission networks. Table 6.15 summarizes the main parameters of an air core reactor for 380 kV. These parameters are used to calculate an AC-SFCL for ultra-high voltage. Firstly, the results for the calculation of a retrofit of an air core reactor are shown and then a low impedance design by variation of the geometry as it is the case for medium and high voltage.

Table 6.15: Main parameters of the air core reactor for ultra-high voltage [Scha13]

Parameter	Symbol	Value
Rated voltage	V_n	380 kV
Rated current	I_r	1 kA
Frequency	f_n	50 Hz
Reference impedance	z_{ref}	6%
Impedance	$\|Z_{ACR}\|$	13.164 Ω
Reactance	X_{ACR}	13.164 Ω
Inductance	L_p	41.901 mH
Resistance (primary winding)	R_p	78 mΩ
Inner diameter	d_{ip}	2004 mm
Outer diameter	d_{ap}	2326 mm
Height	h_p	3093 mm
Number of turns	N_p	199.5
Conductor cross-section	A_{cop}	733 mm²
Fill-factor	f_{fp}	0.2937
Current density	j_p	1.364 A / mm²

Retrofit of an air core reactor

The input parameters for the calculation of the retrofit for ultra-high voltage are summarized in Table 6.16. The value (ranges) are the same as for the high voltage retrofit with exception of winding distance d_w. Due to the ultra-high voltage it was assumed, that more electrical insulation is needed. Therefore the distance between the windings must be increased. Although the number of turns of the primary winding for ultra-high voltage is substantially higher compared to high voltage the value range of the maximum current I_{max} remains unaltered, because more *RE*BCO tapes are needed to effectively shield the increased height at ultra-high voltage.

Table 6.16: Input parameters for calculation of the retrofit

Parameter	Symbol	Value (range)
Maximum current	I_{nax}	200 A - 600 A
Thickness of copper stabilizer	h_{stab}	0 µm – 60 µm
Thickness of silver stabilizer	h_s	2 µm
Winding distance	d_w	5 cm - 15 cm
Tape width	w_t	12 mm
Fill-factor secondary winding	f_{fs}	0.8

The calculated results for impedance during normal operation and during fault operation depending on winding distance d_w and stabilizer thickness h_{stab} for a maximum current I_{max} = 300 A are shown in Figure 6.14.

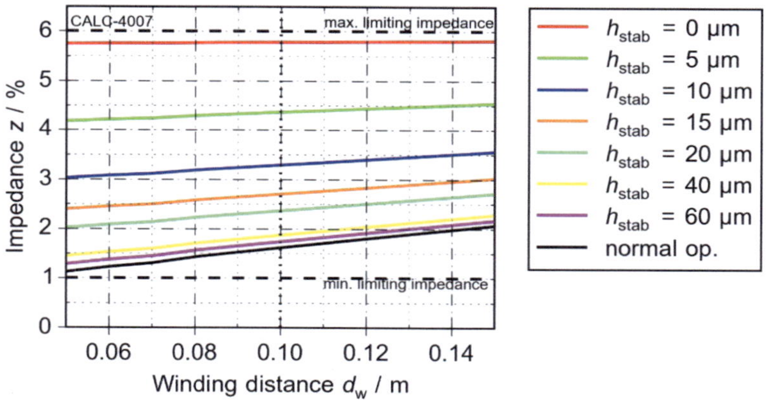

Figure 6.14: Calculated impedance of the retrofit for normal operation and fault condition for constant current I_{max} = 300 A

The impedance during normal operation and during fault is increasing with increasing winding distance d_w. Additional copper stabilization lowers the impedance during fault and as it is the case for the medium and high voltage retrofit, but to a higher degree due to the increased number of *RE*BCO tapes. The impedance of the air core reactor z_{lim} = 6% is not reached even without additional copper stabilization. This is caused by the increased winding distance, which reduces the diameter of the secondary winding and the cross-section of the couple magnetic field. Therefore the main reactance and the impedance are reduced as well. The winding distance of d_w = 10 cm also increases the stray reactance to such an extent, that the $z_n \leq 1\%$ criterion is not reached. For this case the ratio of impedance during fault to impedance during normal operation of Z_{lim}/Z_n = 2.7 with 5 µm of additional copper stabilization.

Figure 6.15 shows the impedance depending on maximum current I_{max} and stabilizer thickness h_{stab} at a constant winding distance of d_w = 10 cm. For any constant stabilizer thickness h_{stab} the impedance during fault is increasing due to the decreasing number of *RE*BCO tapes n_{tot} needed to carry the induced current. The main electrical parameters of the retrofit for ultra-high voltage are summarized in Table 6.17.

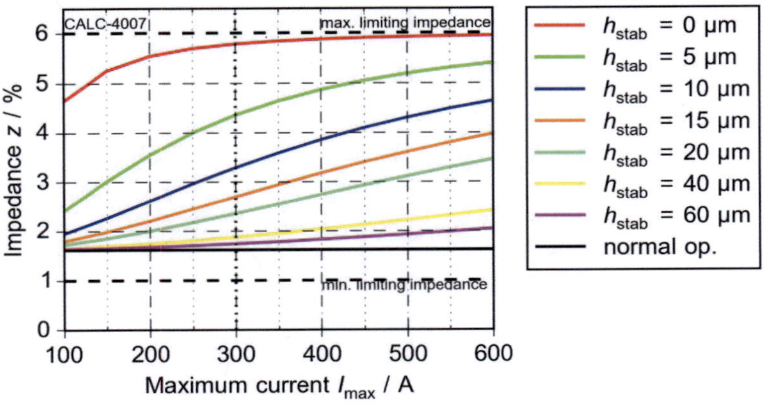

Figure 6.15: Calculated impedance of the retrofit for normal operation and fault condition for constant winding distance d_w = 10 cm

Table 6.17: Main electrical parameters of the calculated retrofit for 380 kV, 1 kA, z = 6%

Parameter	Symbol	Value	p.u.
Primary Resistance	R_p	93 mΩ	0.042%
Primary stray reactance	$X_{\sigma p}$	i2.39 Ω	1.09%
Main reactance	X_m	i10.82 Ω	4.93%
Secondary stray reactance	$X'_{\sigma p}$	i0.029 mΩ	0.013‰
Secondary resistance	R'_s	10.47 Ω	4.77%
Impedance during fault	Z_{lim}	(5.5+7.57i) Ω	4.26%
Impedance during normal operation	Z_n	(0.093+2.39i) Ω	1.09%

During normal operation the impedance is z_n = 1.62% and during fault operation is z_{lim} = 4.37% resulting in ration of z_{lim}/z_n = 2.7. This ratio could be increased by lowering the copper stabilization. However this is not disadvantageous in terms of thermal and electrical stabilization of the *RE*BCO tape.

Low impedance design

The input parameters for the calculation of the low impedance are summarized in Table 6.18. The fill-factor of the primary winding f_{fp} and the conductor cross-section A_{Cu} remained the same as in the case of the air core reactor. The winding distance was set to d_w = 10 cm and no additional copper stabilization was assumed.

Table 6.18: Input parameters for calculation of the low impedance design

Parameter	Symbol	Value (range)
Fill-factor primary winding	f_{fs}	0.2937
Conductor cross-section	A_{Cu}	733 mm
Winding distance	d_w	10 cm
Additional copper stabilization	h_{stab}	0 µm
Inner radius	r_{ip}	1.0 m – 2.0 m
Height	h_p	2.0 m – 4.0 m

The inner radius r_{ip} and height h_p of the primary are varied and in the shown value range and the number of turns calculated for each geometry assuming only silver stabilization criterion (Z_{ACR} = 13.164 Ω). Each calculated design assumes a secondary winding with is highly resistive during fault operation and capable of carrying the induced current during normal operation using one layer of *RE*BCO tapes.

Figure 6.16 shows the calculated results for impedance during normal operation depending on number of turns N_p, inner radius r_{ip} and height h_p.

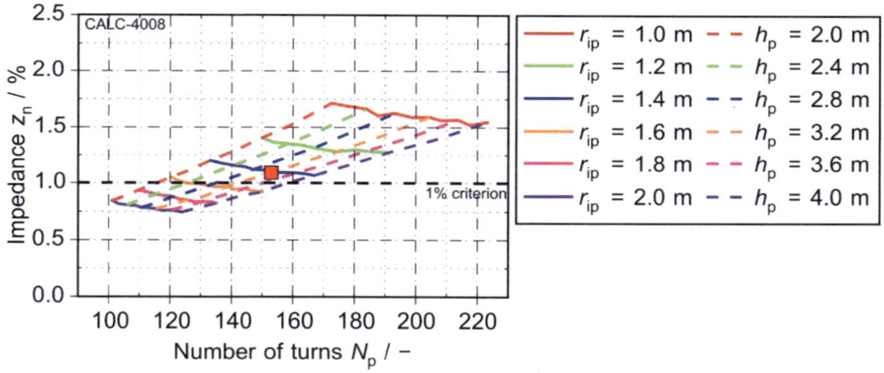

Figure 6.16: Results for impedance during normal operation of 380 kV, 1 kA, z = 6% AC-SFCLs for variation of the geometry of the primary winding. The red square indicates the chosen low impedance design

To meet the $z_n \leq 1\%$ criterion the inner radius must be at least r_{ip} > 1.2 m for any height h_p. Accordingly the height must be at least h_p > 3.2 m for any inner radius r_{ip}. For the low impedance design a geometry with an inner radius r_{ip} = 1.2 m and height h_p = 3.2 m was chosen. For geometry the impedance during normal operation should drop to z_n = 0.8%. For this geometry a secondary winding was calculated using the same input parameters as for the retrofit. Figure 6.17 shows the results for impedance depending on winding distance d_w and stabilizer thickness h_{stab} at a constant maximum current of I_{max} = 300 A. As it is the case for the retrofit the low impedance design does not reach the impedance of the air core reactor of z_{ACR} = 6%. Additional copper stabilization lowers the impedance to a stronger degree compared to medium and high voltage due to the increased number of *RE*BCO tapes n_{tot} as it is the case with the retrofit.

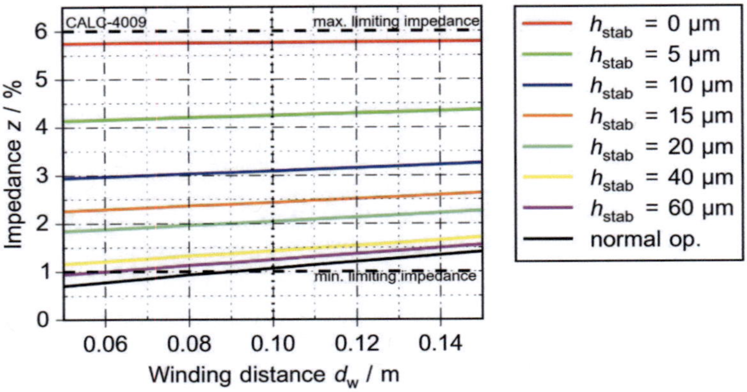

Figure 6.17: Dependency of the calculated impedance on winding distance and stabilization for the low impedance design (I_{max} = 300 A)

For a winding distance d_w = 10 cm the impedance during normal operation is z_n = 1.3%. The respective ratio of impedance during fault operation and impedance during normal operation is z_{lim}/z_n = 3.3 with a maximum copper stabilization of h_{stab} = 5 µm.

Figure 6.18 shows the dependence of the impedance on maximum current I_{max} and copper stabilization h_{stab} for a constant winding distance of d_w = 10 cm.

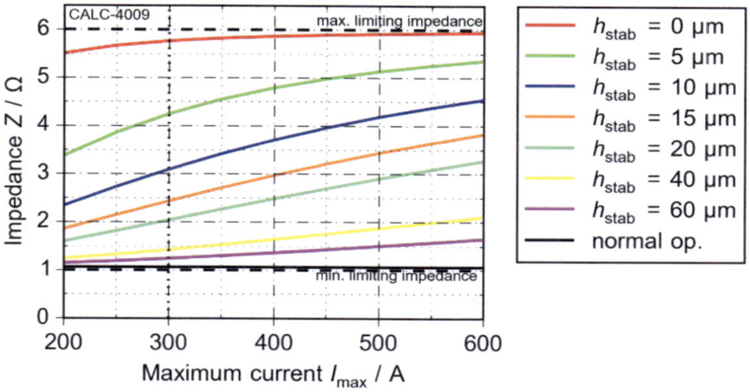

Figure 6.18: Dependency of the calculated impedance on critical current and stabilization for the low impedance design (d_w = 10 cm)

Table 6.19 summarizes the main electric parameters of the low impedance design. The resistance R_p of the primary winding is decreased by 0.76% compared to the retrofit. As expected the primary stray reactance is lower compared to the retrofit.

Table 6.19: Main electrical parameters of the low impedance design

Parameter	Symbol	Value	p.u.
Primary Resistance	R_p	73 mΩ	0.033%
Primary stray reactance	$X_{\sigma p}$	i2.86 Ω	1.3%
Main reactance	X_m	i10.3 Ω	4.69%
Secondary stray reactance	$X'_{\sigma s}$	i0.023 mΩ	0.01‰
Secondary resistance	R'_s	10.23 Ω	4.66%
Impedance during fault	Z_{lim}	(5.23+7.97i) Ω	4.34%
Impedance during normal operation	Z_n	(0.073+2.86i) Ω	1.3%

The main reactance is slightly increased by 0.31% as well as the impedance during fault is slightly lowered by 0.04% compared to the retrofit. The ration of impedance during fault compared to impedance during normal operation is Z_{lim}/Z_n = 3.3. As it is the case for the retrofit the ration Z_{lim}/Z_n can be improved by reducing the copper stabilization.

Comparison of results

The main parameters of the retrofit and the low impedance design are summarized in Table 6.20.

Table 6.20: Comparison of retrofit and low impedance design

Parameter		Symbol	Retrofit	Low imp. design
Height		h_p	3093 mm	3200 mm
Inner diameter		d_{ip}	2004 mm	2800 mm
Outer diameter		d_{ap}	2326 mm	3040 mm
Number of turns		N_p	199.5	153.75
Maximum field at	center	$B_{max\parallel}$	70 mT	28 mT
	end	$B_{max\perp}$	39 mT	34 mT
Number of tapes		n_{tot}	941	725
Maximum current		I_{max}	300 A	300 A
*RE*BCO length		l_{tot}	5336 m	5922 m
		l_{turn}	5.67 m	8.17 m
Stabilizer thickness		h_{stab}	5 µm	5 µm
Total AC-losses		P_{tot}	403.5 W	619.27 W
Impedance	normal op.	Z_n (z_n)	3.55 Ω (1.61%)	2.39 Ω (1.09%)
	fault op.	Z_{lim} (z_{lim})	9.59 Ω (4.37%)	9.36 Ω (4.26%)

The increase of diameter d_{ip} and height h_p of the low impendence winding causes an increase of needed *RE*BCO conductor, but at the same time lowers the occurring maximum magnetic field densities. These lowered magnetic field densities, the lesser number of turns of the primary winding N_p and the increased height lower the burden in terms of maximum current I_{max} of the low impedance design. The copper stabilizer thickness of h_{stab} = 10 µm is the same for both designs. The AC-losses are summarized in Table 6.21 and show a slight advantage for the retrofit (16% less AC-losses). They are mainly determined by the self-field losses, but show a higher contribution of losses due to external perpendicular field compared to the designs for medium and high voltages.

Table 6.21: AC-losses of the retrofit and the low impedance design

Parameter	Retrofit	Low impedance design
External perpendicular field	9.61 W	31.53 W
External parallel field	0.025 W	0.029 W
Self-field	393.8 W	587.4 W
Eddy current	0.103 W	0.303 W
Sum	403.5 W	619.3 W

Figure 6.19 shows the expected fault current limitation of the retrofit and the low impedance design. For calculation of the steady-state fault current limitation eq. (4.3) to eq. (4.6) were used.

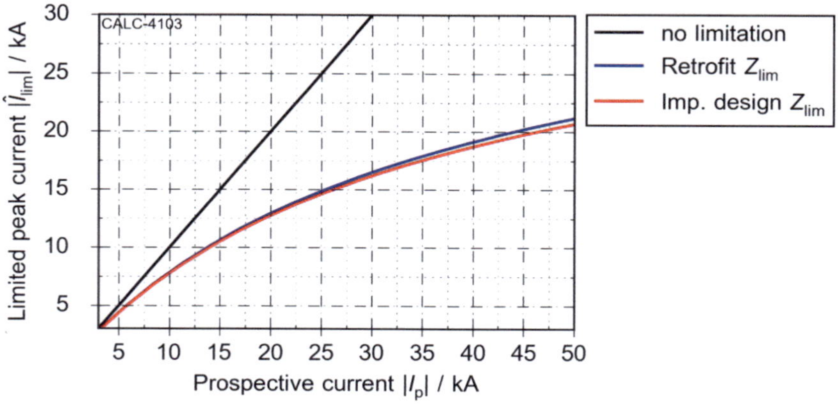

Figure 6.19: Expected fault current limitation of the retrofit and the low impedance design for 380 kV, 1 kA, z = 6%

The small difference of impedance during fault operation z_{lim} of both designs has no effect on the steady-state fault current limitation. Both designs limit a fault current to the same degree. A fault current of I = 20 kA (20·I_n) will be limited by 25% to I_{lim} = 15 kA for example. For both designs the fault current limitation increases with the prospective current.

6.4 Summary

In this chapter conceptual AC-SFCL designs for medium, high and ultra-high voltage have been presented. The calculations based on the equations and design process presented in chapter 1. For each voltage level a retrofit of a typical air core reactor with an impedance of z = 6% was calculated. The objective was achieving a ratio between impedance during fault and impedance during normal operation of at least z_{lim}/z_n = 5 and an impedance during normal operation of $z_n \leq$ 1%.

Additionally low impedance designs for each voltage level have been calculated. The objective of the low impedance designs was to increase the ratio of impedance during fault and impedance during normal operation with only a modest increase in dimension.

The geometry parameters, the major variable parameters and, AC losses and the impedances of all designs (retrofit and low impedance) are summarized in Table 6.22.

For all calculated designs a maximum current of I_{max} = 300 A and a width of w_t = 1.2 cm of the *RE*BCO tape was assumed. Generally the dimensions and the number of turns N_p of the conceptual AC-SFCLs increase with voltage level. Consequently the *RE*BCO demand and AC losses are increasing with voltage level as well.

In case of medium voltage (10 kV) the impedance during normal operation is z_n = 1 with a winding distance of d_w = 20 mm. This matched exactly the $z_n \leq$ 1% criterion. The ratio of impedance during fault and impedance during normal operation is z_{lim}/z_n = 5 with a maximum stabilizer thickness of h_{stab} = 10µm, which as well matches the defined ration of z_{lim}/z_n = 5. The low impedance design for medium voltage reduces the impedance during normal operation even further to z_n = 0.66%. Because of the increased dimensions and the same number of turns N_p of the low impedance design the inductance, and the impedance accordingly, increases as well using the same stabilizer thickness. The result is an increased impedance during fault operation and an increased impedance ratio of z_{lim}/z_n = 6%

Table 6.22: Major data of calculated conceptual designs

Parameter	10 kV, 1 kA		110 kV, 1 kA		380 kV, 1 kA	
	Retrofit	Low impedance	Retrofit	Low impedance	Retrofit	Low impedance
Inner diameter d_i	1101 mm	1200 mm	1854 mm	2000 mm	2004 mm	2800 mm
Outer diameter d_a	1184 mm	1267 mm	2160 mm	2298 mm	2326 mm	3040 mm
Height h_p	990 mm	1200 mm	1210 mm	1200 mm	3093 mm	3200 mm
Number of turns N_p	36.5	36.9	84.75	81.86	199.5	153.75
Winding distance d_w	20 mm	20 mm	30 mm	30 mm	100 mm	100 mm
Stabilizer thickness h_{stab}	10 µm	10 µm	5 µm	5 µm	5 µm	5 µm
SC length	579 m 3.34 m/turn	646 m 3.64 m/turn	1196 m 5.64m/turn	1428 m 6.1 m/turn	5336 m 5.67 m/turn	5922 m 8.17 m/turn
AC losses	73.94 W	99.05 W	166.3 W	170.2 W	403.5 W	619.27 W
Impedance z_n	1% (59.3 mΩ)	0.66% (38.1 mΩ)	1.11% (0.707 Ω)	0.84% 4.74 mΩ	1.61% (3.55 Ω)	1.09% (2.39 Ω)
Impedance z_{lim}	5.07% (293 mΩ)	5.47% (316 mΩ)	4.81% (3.05 Ω)	4.74% (3.01 Ω)	4.37% (9.59%)	4.26% (9.36 Ω)

Compared to medium voltage, the AC-SFCL for high voltage (110 kV) the geometry increases mainly in diameter. The number of turns are increasing as well in order to achieve the 6% of impedance of the air core reactor for this voltage level. With an as well increased winding distance of d_w = 30 mm, the impedance during normal operation is z_n = 1.11%, which is slightly above the defined $z_n \leq$ 1%. The impedance during fault reaches z_{lim} = 4.81%, which results in an impedance ratio of z_{lim}/z_n = 4.3%. This is lower than requested. The low impedance design improves therefore the retrofit with respect to the impedances. The increase in diameter, almost the same height and fewer number of turns N_p reduce the impedance during normal operation to z_n = 0.84% at the same winding distance of d_w = 40 mm. Assuming the same stabilizer thickness of h_{stab} = 5 µm the impedance is slightly lower compared to the retrofit, the impedance ratio however is increasing to z_{lim}/z_n = 5.6.

For ultra-high voltage (380 kV) the dimension and number of turns increase again. For the retrofit the diameter is almost the same, but the height is tremendously increased compared to high voltage. For sufficient electric insulation the winding distance needs to be increased for this voltage level. As consequence the impedance during normal operation is z_n = 1.61%, which is above the $z_n \leq 1\%$ criterion. Allowing the same stabilizer thickness of h_{stab} = 5 µm the impedance during fault is z_{lim} = 4.37%, which gives an impedance ratio of z_{lim}/z_n = 2.7. Increasing the diameter for the low impedance design, reduces the necessary number of turns N_p in order to achieve the same inductance, and therefore impedance, as the retrofit. As a consequence the cross-section of the primary winding is smaller and accordingly the stray inductance over the primary winding. This means, that the impedance during normal operation is reduced and slightly above the $z_n \leq 1\%$ criterion with z_n = 1.09%. The impedance during fault of the low impedance design is similar to the impedance during fault of the retrofit. This means that the impedance ratio is increasing to z_{lim}/z_n = 3.9.

In conclusion the conceptual designs for medium voltage (10 kV) show, that an AC-SFCL is feasible with low impedance during normal operation ($z_n \leq 1\%$) and a significant impedance increase during fault limitation of at least z_{lim}/z_n = 5. At high voltage (110 kV) this becomes more difficult and the retrofit does not comply with the defined criterions for impedance, even with a smaller winding distance d_w. Therefore the geometry of the primary winding must be adjusted, which is the case for the low impedance design. A modest increase of dimensions enables an impedance z_n <1% and an impedance ratio of z_{lim}/z_n > 5. For ultra-high voltage (380 kV) the challenge of low impedance during normal operation and high impedance during fault becomes even more evident. The retrofit for ultra-high voltage does not comply with both impedance criterions. Compared to medium and high voltage the increase in diameter for the low impedance design is rather high in order to minimize the number of turns N_p and therefore the stray inductance, whilst maintaining the same impedance. With this measure it is possible to reduce the impedance during normal operation to next to z_n = 1%. However to fully meet both impedance criterions it is necessary to further adapt the geometry. As shown in Figure 6.7, Figure 6.13, Figure 6.19 all designs exhibit the same current limitation behavior: Increasing current limitation with increasing prospective current. The current limitation is hereby independent of the voltage level and only dependent on the impedance during fault z_{lim}, which is in the range of 4% to 5.5% for all designs.

7 Summary and conclusions

Reliable and effective limitation of fault currents in power grids is a crucial and demanding task to maintain a non-interruptible supply of electric energy and prevent damage of electrical devices in the power grid. Conventional and commercial solutions to limit fault currents are implemented, but have disadvantages. Fuses can be used for a single event and need to be replaced after fault current limitation. The replacement can be time consuming and prolongs the power outage. Air core reactors do have a non-negligible impedance during normal operation, which can cause instabilities and system perturbations.

Superconducting fault current limiters (SFCLs) have been field tested and are on the verge of becoming a commercial and accepted measure to limit fault currents. In general, SFCLs offer fast (within the first half cycle) and reliable current limitation, while maintaining a low impedance during normal operation and can recover under load. Several different types of SFCLs are subject to research and development [Mor13], [EPRI09], [NS07].

This work proposes and investigates the Air Core Superconducting Limiter (AC-SFCL) as a future measure to protect the power grid from fault currents. The initial objective of the AC-SFCL is to improve the air core reactor by retrofitting it with a secondary superconducting winding. The secondary superconducting winding shields the primary winding and lowers the reactance, and therefore the impedance, of the air core reactor significantly in normal operation mode. In fault-limiting operation the secondary winding becomes normal conducting, the resistance increases and triggers the main inductance and therefore increases the impedance during faults significantly.

To design an AC-SFCL all necessary equations were derived and formulated to connect the geometry and properties of the *RE*BCO tapes with the electrical parameters. The equations were integrated in a design method to automate the calculation and variation of the main parameters. Hereby the design process was divided into two sub-processes depending on the application. One application is the retrofit of a conventional air core reactor, the other is to optimize the primary winding, essentially adapting the air core reactor geometry to enable an even lower impedance during normal operation.

With the equations and design method a $S = 60$ kVA, $V = 400$ V and $z = 6\%$ AC-SFCL demonstrator was designed and the current limiting capability estimated using a steady-state approach for the impedance during fault. A major design objective was to achieve an impedance $z < 1\%$ during normal operation, while keeping the geometry most compact. Simulations showed an increasing fault current limitation with increasing fault current and a maximum fault current imitation 40% at $I_p = 3.6$ kA, while the impedance under load conditions should be at $z_n = 0.68\%$.

Before building the AC-SFCL demonstrator the *RE*BCO tapes have been characterized and tested for suitability for the application. A small test setup served as blueprint to investigate the manufacturing process, the quench behavior of single superconducting *RE*BCO rings and the test bed for short-circuit tests.

The primary winding of the AC-SFCL demonstrator was manufactured using copper wire with lacquer insulation onto a G10 bobbin. Reinforcements added within the bobbin during the winding process allowed to reduce the wall thickness and therefore the overall dimensions of the AC-SFCL as well as the distance of the primary and secondary winding. The secondary winding was manufactured aligning 22 single, short-circuited *RE*BCO rings with 12 mm tape width and copper stabilization.

For the load and short-circuit tests both windings were assembled and fully emerged into liquid nitrogen in an open bath cryostat at ambient pressure. For the load test (normal operation) the impedance was measured as z_n = 0.72%. This is even below the specified z_n < 1% and in very good agreement with the calculated z_n = 0.68%. Short-circuit tests were performed for prospective currents from I_p = 0.65 kA to I_p = 3.6 kA and showed the expected increasing short-circuit limitation with increasing short-circuit to the same degree as the steady-state simulation. The fault current limitation was found to be lower in the first half cycle compared to the following half cycles. Calculation of the phase angle showed, that the inductance of the AC-SFCL demonstrator is not fully developed in the first half cycle. With beginning of the second half cycle the inductance is fully developed and the measured current limitation is in good agreement with the steady-state simulation, showing the best current limitation of 40% at a prospective current of I_p = 3.6 kA. At a prospective current of I_p = 1.0 kA the resistance reaches its design value and stays within 5% for higher fault currents and at I_p = 1.35 kA the limitation starts to be more inductive than resistive. The AC-SFCL demonstrator maintains this fault current limitation regardless of the phase angle of the fault.

The results prove the concept of the AC-SFCL and testify a reliable and fast fault current limitation. The measurements are in good agreement with the simulation and with the calculated design values proving the design process and equations as valid.

Further testing of fault current limitation has been performed with a power hardware in-the-loop system. Three interface algorithms have been investigated in order to ensure a stable and reliable operation of the PHIL system and the connected AC-SFCL demonstrator. A modified damping interface algorithm showed the best results in terms of peak current and implementing this algorithm the results for fault current measurement could be reproduced successfully. The PHIL setup allowed to simulate a three phase system in which the physical AC-SFCL with the power hardware was connected to one phase A and phase B and C contained virtual AC-SFCLS. In each phase a load and a short-circuit path was inserted and all three phases were connected to a transient model of a synchronous generator. With this setup symmetrical and unsymmetrical short-circuits have been simulated. The AC-SFCL demonstrator was capable of limiting any occurring fault current in symmetrical and unsymmetrical short-circuits to the same degree as in the single phase measurements.

Using the verified design equations and design method conceptual designs for medium, high voltage and ultra-high voltage have been calculated. For all voltage levels designs have been presented, which show a significant increase from impedance during normal operation to impedance during fault. The designs for high and ultra-high voltage show, that it is beneficial to adjust the dimensions of the primary winding of typical air core reactors. This lowers the impedance during normal operation and increases the ratio of impedance during fault to impedance during normal operation accordingly. An increase of dimensions for these low impedance designs comes at the expense of higher conductor demand and AC losses. Nevertheless this increase is rather modest for medium and high voltage and acceptable for ultra-high voltage. The calculation of the steady-state fault current limitation of the conceptual designs show an effective current limitation and increasing current limitation with increasing prospective current as it has been shown with the demonstrator. Generally the results are promising and show, that the AC-SFCL can be an excellent measure to limit fault currents, which can be adapted easily to different voltage levels.

The results presented and experienced gained in this work can be seen as a solid base for future developments. The major prospect is the design, building and testing of an AC-SFCL demonstrator, which proves the feasibility of the AC-SFCL concept for medium voltage and further investigation of the manufacturing process. Emphasize should hereby be given on the thermal and electrical insulation of the secondary winding in order to minimize the distance of the windings and therefore the impedance during normal operation. Another challenge, which could be rather easily addressed is the further improvement of the soldering for the *RE*BCO rings.

Further investigation of the quench behavior of *RE*BCO rings or windings of stacked *RE*BCO rings would be useful to optimize the secondary winding of the AC-SFCL. Hereby *RE*BCO tapes with different stabilization or different critical current $I_c(B,T)$ could be qualified for different positions in the secondary winding, e.g. at the end or in the center.

Power hardware-in-the-loop systems could be used to further investigate the operational behavior of the AC-SFCL under different grid conditions, e.g. bus bar coupling, and show the advantage of the low impedance compared to the air core reactor in terms of grid stability and system perturbations.

A. Additional Information on devices and resources used for this work

Characterization of *RE*BCO tapes

R(*T*) measurement

The measurement setup for the *R*(*T*) measurement is shown in Figure 4.1. The following devices have been used for this measurement:

Temperature sensor	PT100 platinum sensor manufactured in thin film technology
Temperature measurement	Lakeshore 218 temperature monitor
Voltage measurement	Hewlett Packard 3458 A multimeter
Power source	Keithley 6221 DC and AC power source
Data acquisition and control	National Instruments LabVIEW 8.0 on Windows PC

$I_c(B,T)$ measurement

Magnets	Bruker B-E15, max. 600 mT
Power sources (Magnet)	Bruker B-H11D and B-MNC5
Hall probe	Arepoc LHP-NP 203, 29.5 mV/T
Power source (hall probe)	Burster-Gernsbach Präzissionsmesstechnik, Präzisionsstromgeber Typ 6426
Voltage measurement (hall probe)	Keithley 2000 multimeter
Voltage measurement (*RE*BCO tape)	Keithley 2182 A nanovoltmeter
Power sources (measurement current)	Agilent 6672 A, 0-20 V / 0-100 A DC power supply Agilent 6681 A, 0-8 V / 0-580 A DC power supply
Data acquisition and control	National Instruments LabVIEW 8.0 on Windows PC

Measurement of critical current at self-field

The setup for measuring the critical current I_c at self-field is shown in Figure 4.7. The following devices have been used for this measurement:

Multiplexer	Agilent 44970 A, 20 channel multiplexer
Voltage measurement	Keithley 2182 A nanovoltmeter
Power sources	Agilent 6672 A, 0-20 V / 0-100 A DC power supply Agilent 6681 A, 0-8 V / 0-580 A DC power supply
Data acquisition and control	National Instruments LabVIEW 2012 on Windows PC

All devices have been connected via a GPIB bus.

Preliminary Measurements

Measurement of resistance of soldering

The following devices have been used for this measurement:

Multiplexer	Agilent 44970 A, 20 channel multiplexer
Voltage measurement	Keithley 2182 A nanovoltmeter
Power sources	Agilent 6672 A, 0-20 V / 0-100 A DC power supply Agilent 6681 A, 0-8 V / 0-580 A DC power supply
Data acquisition and control	National Instruments LabVIEW 2012 on Windows PC

All devices have been connected via a GPIB bus.

Measurement of inductance of coils

LCR-Meter	Instek LCR-821

Quenching single superconducting rings

The measurement setup for quenching single superconducting rings is shown in Figure 4.15. The following devices have been used for this measurement:

Power source	SBA Sn = 400 kVA, 50 Hz, V_{prim} = 225 V / 400 V adjustable, V_{sec} = 50-1000 V adjustable in 50 V steps, $I_{prim,max}$ = 200 A continuously, $I_{sec,max}$ = 400 A continuously Primary side set to V_{prim} =230 V Secondary side set to V_{sec} = 50 V
Thyristor (load)	GvA Leistungselektronik GmbH, AC-switch W1C 250 V, 10 kA SE/EB001, Snubber circuit: R = 2.2 Ω, C = 3 µF
Thyristor (fault)	GvA Leistungselektronik GmbH, short-circuit switch 1.5 kV, 10 kA 5STB18N4200, Snubber circuit: R = 10 Ω, C = 6.6 µF
Resistance (load)	Heine Dresden, Power resistors 0.3-1 Ω, 750-2500 A
Resistance (fault)	Customized design 0-800 mΩ continuously adjustable
Rogowski coil	Rocoil SE 432 used with three channel integrator
Differential probe	Tektronics P5200 high voltage differential probe, 1:50 or 1:500. Set to 1:50
Transient recorder	Elsys AG, 16 Channels, max. Voltage V_{max} = 10 V
Data acquisition and control	TransAS 3.0

Testing the AC-SFCL demonstrator

The measurement setup for the test with the AC-SFCL demonstrator is shown in Figure 4.25. The following devices have been used for this setup:

Power source	SBA Sn = 400 kVA, 50 Hz, V_{prim} = 225 V / 400 V adjustable, V_{sec} = 50-1000 V adjustable in 50 V steps, $I_{prim,max}$ = 200 A continuously, $I_{sec,max}$ = 400 A continuously Primary side set to V_{prim} =400 V Secondary side set to V_{sec} = 400 V
Thyristor (load)	GvA Leistungselektronik GmbH, AC-switch W1C 250 V, 10 kA SE/EB001, Snubber circuit: R = 2.2 Ω, C = 3 µF
Thyristor (fault)	GvA Leistungselektronik GmbH, short-circuit switch 1.5 kV, 10 kA 5STB18N4200, Snubber circuit: R = 10 Ω, C = 6.6 µF
Resistance (load)	Heine Dresden, Power resistors 0.3-1 Ω, 750-2500 A
Resistance (fault)	Customized design 0-800 mΩ continuously adjustable
Rogowski coil	Rocoil SE 432 used with three channel integrator
Differential probe	Tektronics P5200 high voltage differential probe, 1:50 or 1:500. Set to 1:50
Transient recorder	Elsys AG, 16 Channels, max. Voltage V_{max} = 10 V
Data acquisition and control	TransAS 3.0

Software

Calculations	MathCAD 14, MatLAB 2012 and the python Numpy 1.9.1 package
Circuit Diagrams, schematics and arrangement for figures	Inkscape 0.48 and 0.91
3D graphics	Blender 2.72
2D engineering drawings	LibreCAD 2.01 and later
Post-processing images	The Gimp 2.8
Processing and plotting measured and calculated data	Python Matplotlib 1.4.2 and python Pandas 0.15.2 packages
Writing	Microsoft Word 2010

B. Calculation of the inductance of an solenoid with arbitrary cross-section

The general approach to calculate the inductance of a solenoid with arbitrary cross-section is shown in Figure B.1. In step 1 the geometry of the solenoid must be given. This includes the inner radius r_i the outer radius r_a and the height h of the solenoid. This geometry is then divided into a finite number of current loops as indicated in Figure 3.6 (step 2). The number of finite loops or the diameter d_{loop} respectively is freely selectable. A higher density of current loops delivers higher accuracy of the calculation, while fewer current loops might speed up the calculation time. Practically diameters of $d_{loop} \approx 10$ mm for air core reactors with dimensions ~ 1 m and $d_{loop} = t_{sc}$, with t_{sc} being the thickness of a *RE*BCO tape, for superconducting windings as used in the AC-SFCL have been found to be sufficiently accurate.

In step 3 the calculation starts by choosing a reference loop and calculate its position within the cross-section (step 4). The same is done with the target loop (step 5 and step 6). This approach reflects the correlation given in eq. (3.41) and eq. (3.42), that the inductance of the solenoid is the sum of all mutual inductances and self-inductances of the current loops.

In step 7 is decided, if the reference loop and the target loop are the same and the self-inductance is calculated (step 9). If this is not the case the geometrical distances in axial and radial direction must be calculated first in order to calculate the mutual inductance of both loops (step 8 and step 9). In step 10 the calculated inductance is added to the total inductance (compare eq. 3.42).

The calculation is continued until the last current loop is processed (steps 11 and 12). At the end the total inductance or sum of all self-inductances and mutual inductances respectively is divided by the square number of current loops in order to calculate the inductance of the winding for one turn ($N = 1$).

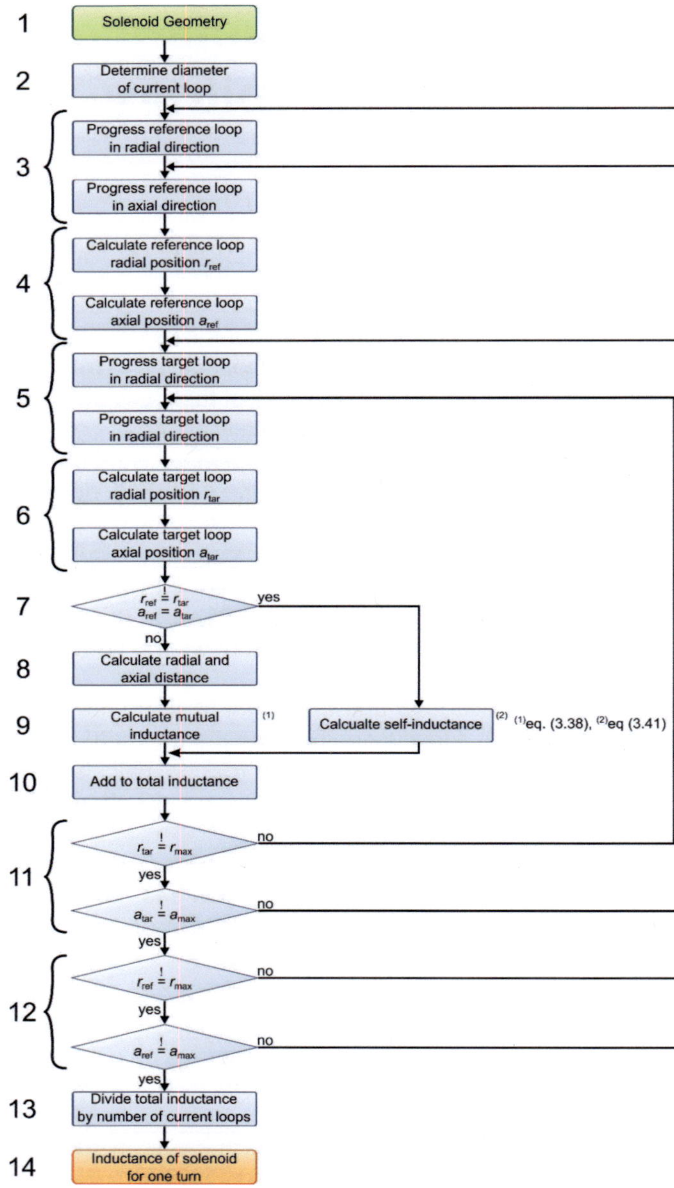

Figure B.1: Flow diagram of the function to calculate the inductance of a solenoid with arbitrary cross section

C. Calculation of phase angle of sine signals at 50 Hz

The impedance of the AC-SFCL can be separated into a resistive and inductive component. In order to extract these components it is necessary to calculate the phase angle φ between voltage and current. In the case of the AC-SFCL this is very important, since the resistance and the inductance are changing significantly during fault operation and therefore the calculation of the phase angle φ allows to determine, which component – resistive or inductive – contributes to the fault current limitation. Additionally this information allows comparing the measurements with the calculated steady-state values.

In order to verify the approach described in chapter 4.4.2 it was tested against simulated sine signals with a given phase shift. In a first step sine signals for voltage and current are generated. The amplitudes of the signals were calculated based on reasonable values for the impedance of an AC-SFCL:

- Impedance for normal operation z_n = 1%
- Impedance during fault limitation z_{lim} = 6%

For voltage and current the specified values of the AC-SFCL demonstrator are used:

- Rated current I_r = 150 A
- Rated voltage V_r = 400 V

Using eq. (2.5) the respective voltage drop over the impedance is

- Normal operation V_{low} = 2.8 V
- Fault operation: V_{high} = 24 V

The phase angle between voltage and current was investigated for three cases:

(A) 0° degrees

(B) 45° degrees

(C) 90° degrees

The generated signals are shown in Figure C.1. The signal length is 10 cycles in all three cases and a sample rate of 100 samples per millisecond. In all cases the current shows the expected peak of I_{peak} = 212 A corresponding to I_r =150 A and the voltage the expected peak of V_{peak} = 3.96 V corresponding to V_r = 2.8 V.

In order to calculate the phase angle for each of the three cases shown in Figure C.1 the hilbert transform function of the scipy software package is applied. This function basically returns a vector of complex numbers with constant magnitude and a constant phase change for the voltage and current signals. Applying the angle function of the numpy software package returns the phase angle for voltage and current. The phase angle is then calculated by subtracting the calculated angle of the voltage and current vectors.

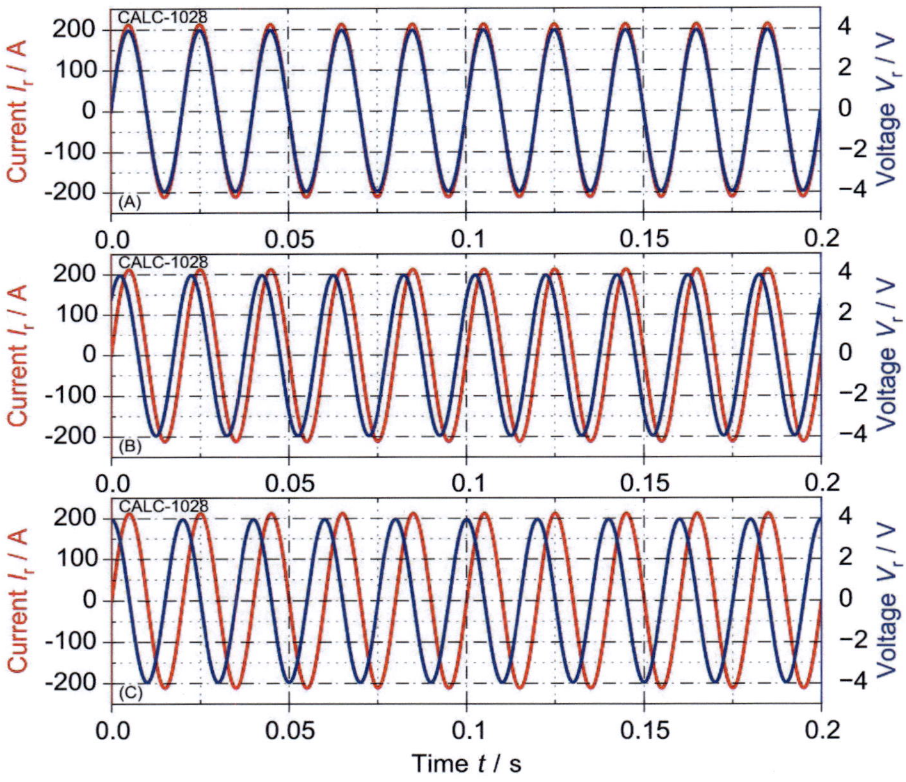

Figure C.1: Generated sine signals for voltage and current with phase angles of (A) 0°, (B) 45° and (C) 90°

The calculated phase angles for the signals in Figure C.1 are shown in Figure C.2. In each case the calculation of the phase angles is very accurate and shows no boundary effects.

During testing the AC-SFCL demonstrator is subject to load and fault currents. This means, that the amplitudes and the phase angle of the measured current and voltage might change rapidly. The reason for this is the change in resistance and mainly inductance of the AC-SFCL demonstrator, when switching from normal operation to fault current limitation.

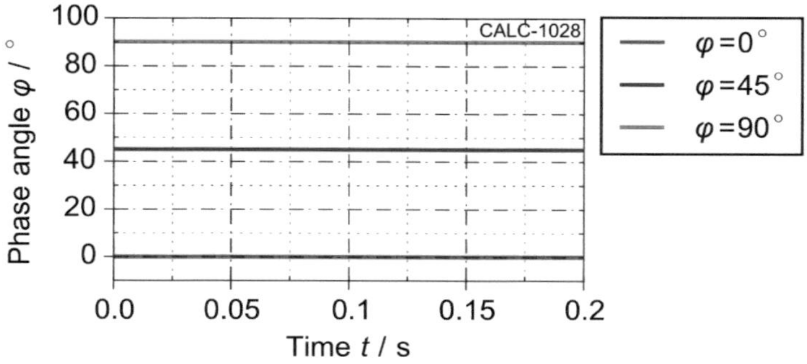

Figure C.2: Respective calculated phase angles φ for signals shown in Figure C.1

In order to verify the described calculation of the phase angle φ for this case voltage and current signals heave been generated with load cycles for 400 ms followed by fault cycles for 200 ms and followed by fault cycles for 400 ms again. The peak current during the fault cycles is set to I_p = 600 A and the voltage drop is V_{high} = 24 V.

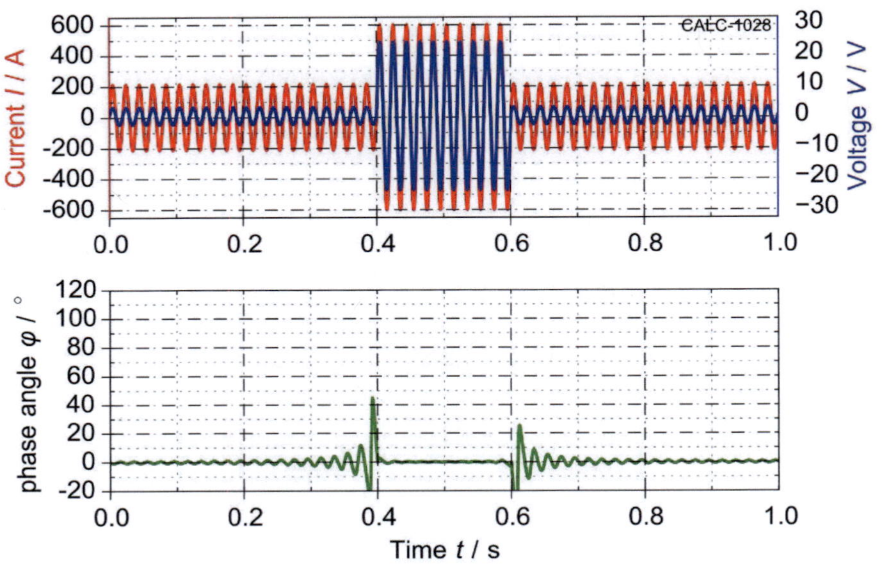

Figure C.3: Load-fault-load cycles. The phase angle between voltage and current is φ = 0° during load and φ = 0° during fault

Figure C.3 show the generated load-fault-load cycles with the calculated phase angle. Hereby the load lasts for 400 ms, then fault operation is assumed for 200 ms and afterwards load operation again for 400 ms. The phase angle was set to φ = 0° for the load cycles and the fault cycles. The calculated phase angle φ from the voltage and the current in Figure C.3 shows minor ripples at the beginning and end of the signals. The ripples are increasing for load cycles the shorter distance is to the fault cycles. At the end of the first load period (t < 400 ms) the strongest ripples occur. After entering the fault operation the ripples almost disappear instantaneously and the phase angle φ during the fault cycles is almost constant. The calculated phase angle during the load cycles starting at t = 600 ms show ripples with decreasing peaks.

Figure C.4 shows the current and voltage signals with load-fault-load periods and the same durations as in Figure C.3. In this case however, the current and the voltage have a phase angle of φ = 45° during the fault cycles.

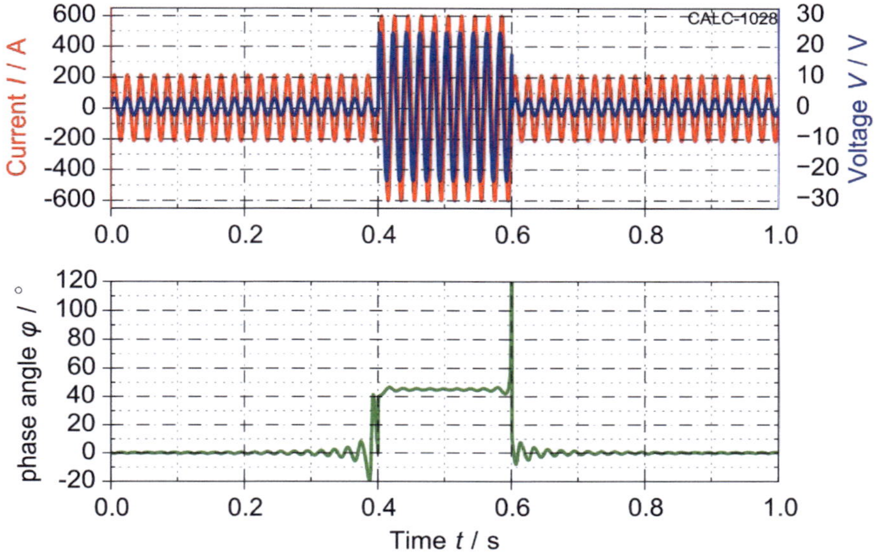

Figure C.4: Load-fault-load cycles. The phase angle between voltage and current is $\varphi = 0°$ during load and $\varphi = 45°$ during fault

The calculated phase angle φ in Figure C.4 shows as well ripples with increasing peaks during the first load period. However, the peaks of these ripples are lower compared to Figure C.3. Before the fault cycles start at $t = 400$ ms the calculated phase angle shows a strong peak, before it settles at $\varphi = 45°$ during the fault cycles. Another strong peak is occurring at $t = 600$ ms, the end of the fault period. Starting with the load cycles at $t = 600$ ms the calculated phase angle shows ripples with decreasing peaks.

Figure C.5 shows again the load-fault-load cycles of current and voltage. In this case the current and voltage have a phase angle of $\varphi = 90°$ during the fault cycles.

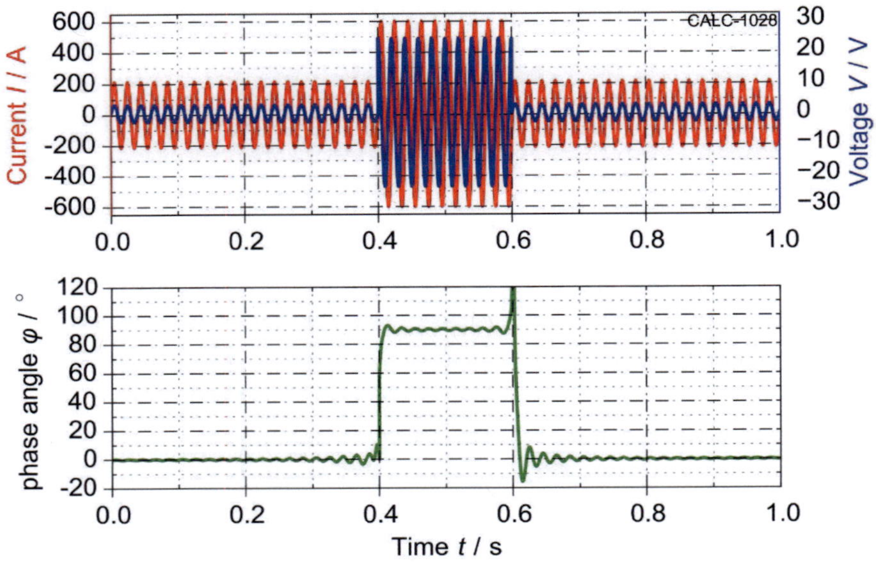

Figure C.5: Load-fault-load cycles. The phase angle between voltage and current is φ = 0° during load and φ = 90° during fault

The calculated phase angle φ in Figure C.5 shows increasing ripples until t = 400 ms during the load cycles, but to a lower degree as in Figure C.3 and Figure C.4. Staring at t = 400 ms the calculated phase angle almost immediately switches to φ = 90°. During the fault cycles ripples are present and at the end a strong peak occurs, before the load cycles are starting. The phase angles of the load cycles at t > 600 ms is similar to the one in Figure C.4.

The calculations show, that the approach to calculate the phase angle φ based on the Hilbert transformation is not suitable for fast changing or transient regimes. However for signals with constant amplitude and frequency the calculation shows a very high accuracy as well as after a very fast change in amplitude and frequency, which is the case for the AC-SFCL.

D. Designations and Abbreviations

a	outer
AC	Alternating current
ACR	Air Core Reactor
AC-SFCL	Air Core Superconducting Fault Current Limiter
Ag	silver
c	critical
CAPS	Center for Advanced Power Systems
CDIM	Classical damping impedance method
Cu	copper
DC	Direct current
DFT	Discrete Fourier Transformation
e	engineering
fb	Feedback
flt	Fault
FSU	Florida State University
G10	Fiberglass reinforced plastic
gg	Generator to ground
hc	Half cycle
i	Inner
IFM	Impedance feedback method
ITEP	Institute for Technical Physics
ITM	Ideal transformer method
KIT	Karlsruhe Institute of Technology
Lg	Line to ground
lim	limitation
LN2	Liquid nitrogen
n	Normal operation

mag	magnitude
max	Maximal
MDIM	Modified damping impedance method
meas	measurement
min	Minimal
nc	Normal conducting
p	primary
par	parallel
PC	Personal Computer
pen	penetration
PHIL	Power Hardware in the Loop
rec	recovery
*RE*BCO	Rare Earth Barium Copper Oxide
RRR	Resistive Residual Ratio
RT	Room temperature
s	secondary
SB	Synchronization Box
sc	superconducting
sfcl	Superconducting fault current limiter
Sim	simulation
sol	solenoid
src	source
stab	stabilization
tot	total
TR	Transient Recorder

E. Index of Symbols

A	Cross-section (A/mm²)
A_{cop}	Cross-section of the conductor in the primary winding (A/mm²)
A_{Cu}	Cross-section of copper wire (A/mm²)
A_{p}	Cross-section of the primary winding (A/mm²)
A_{pen}	Surface penetrated by a magnetic field of a superconductor (mm²)
A_{s}	Cross-section of the secondary winding (A/mm²)
A_{tape}	Cross-section of a *RE*BCO tape (A/mm²)
B	Magnetic flux density (T)
B_{c1}	First critical magnetic flux density (T)
B_{c2}	Second critical magnetic flux density (T)
B_{p}	Magnetic field density necessary to fully penetrate a superconductor (T)
$B_{\parallel}$	Magnetic field density parallel to a superconductor (T)
$B_{\perp}$	Magnetic field density perpendicular to a superconductor (T)
B_{r}	Radial magnetic flux density of a solenoid winding (T)
B_{z}	Parallel magnetic flux density in a solenoid winding (T)
b_{sc}	Width of the superconducting tape (mm)
c_{f}	Coupling factor between two solenoid windings
d_{ap}	Outer diameter of the primary winding (m)
d_{ip}	Inner diameter of the primary winding (m)
d_{w}	Distance between primary and secondary winding (cm)
E	Electrical field (V/cm)
E_{c}	Critical electrical field (V/cm)
E_{m}	Energy stored in a magnetic field (J)
f_{fp}	Fill-factor of the primary winding
f_{fs}	Fill-factor of the secondary winding
f_{n}	Frequency (Hz)
f_{z}	Ratio of impedance during fault and impedance during normal operation

H	Magnetic field strength (A/m)
H_m	Magnetic field strength between the windings of the AC-SFCL (A/m)
H_{demo}	Total height of the ACSFCL demonstrator
h_p	Height of the primary winding (m)
h_s	Height of the secondary winding (m)
h_{stab}	Thickness of copper stabilization of *RE*BCO tape (m)
I	Current (A)
I_c	Critical Current (A)
I_k''	
I_{lim}	Limited fault current (A)
I_{max}	Maximum current allowed in one *RE*BCO tape (A)
I_n	Nominal Current (A)
I_p	Prospective current (A)
I_{phaseB}	Simulated current in phase B in a PHIL system (A)
I_{phaseC}	Simulated current in phase C in a PHIL system (A)
I_{sim}	Simulated Current in one phase PHIL system (A)
j	Current density (A/mm²)
j_c	Critical current density (A/mm²)
j_e	Engineering current density (A/mm²)
j_{ep}	Engineering current density of the primary winding (A/mm²)
j_{es}	Engineering current density of the secondary winding (A/mm²)
$j_{p,con}$	Current density of the conductor in the primary winding (A/mm²)
l_{tot}	Total length of *RE*BCO tape required for a winding (m)
l_{turn}	Length of one turn of a *RE*BCO tape in a winding (m)
L_m	Main inductance (H)
L_{mp}	Primary main inductance (H)
L_{ms}	Secondary main inductance (H)
L_σ	Total stray inductance (H)
$L_{\sigma p}$	Primary stray inductance (H)
$L_{\sigma s}$	Secondary stray inductance (H)

$L'_{\sigma s}$	Secondary stray inductance (referred to primary side) (H)
L_p	Inductance of the primary winding (H)
$L_{p,ref}$	Design inductance of the primary winding (H)
L_s	Inductance of the secondary winding (H)
L_{sol}	Inductance of a solenoid with one winding (H)
M	Mutual inductance (H)
N	Number of turns of a solenoid winding
N_{max}	Maximum number of turns of a solenoid winding
N_{min}	Minimum number of turns of a solenoid winding
N_p	Number of turns of the primary winding
N_s	Number of turns of the secondary winding
n_{par}	Number of parallel *RE*BCO tapes in the secondary winding
n_{tot}	Total number of *RE*BCO tapes in the secondary winding
P_{ac}	Total AC losses (W)
P_e	AC losses due to Eddy Currents (W)
$P_{he\parallel}$	AC losses due to external parallel magnetic field (W)
$P_{he\perp}$	AC losses due to external perpendicular magnetic field (W)
P_{hs}	AC losses due to self-field (W)
r_{ap}	Outer radius primary winding (m)
r_{ip}	Inner radius primary winding (m)
r_{as}	Outer radius secondary winding (m)
r_{is}	Inner radius secondary winding (m)
R_{gg}	Generator to ground resistance in PHIL simulation (Ω)
R_{lg}	Line to ground resistance in PHIL simulation (Ω)
R_{fb}	Feedback resistance in PHIL systems (Ω)
$R_{flt\text{-}A}$	Resistance to adjust fault current in phase A in PHIL simulation (Ω)
$R_{flt\text{-}B}$	Resistance to adjust fault current in phase B in PHIL simulation (Ω)
$R_{flt\text{-}C}$	Resistance to adjust fault current in phase C in PHIL simulation (Ω)
R_n	Resistance of the AC-SFCL during normal operation (Ω)
R_p	Resistance primary winding (Ω)

R_{par}	Parallel resistance (Ω)
R_{s}	Resistance of the soldered connection of two tapes (Ω)
R_{sc}	Resistance of the superconductor in the secondary winding (Ω)
R'_{sc}	Resistance of the superconductor in the secondary winding referred to the primary side (Ω)
R_{sfcl}	Resistance of the AC-SFCL (Ω)
R_{stab}	Resistance of the stabilization of the *RE*BCO tapes in the secondary winding (Ω)
R'_{stab}	Resistance of the stabilization of the *RE*BCO tapes in the secondary winding referred to the primary side (Ω)
R_{src}	Resistance of a power source (Ω)
R_{var}	Variable resistance (Ω)
S_{ACR}	Apparent power of the air core reactor (VA)
S_{base}	Apparent power of a generator in PHIL tests (VA)
S_{n}	Nominal apparent Power (VA)
S_{sfcl}	Apparent power of the AC-SFCL (VA)
t	Time (s)
t_{rec}	Recovery time (s)
T	Temperature (K)
T_{c}	Critical Temperature (K)
V_{n}	Nominal voltage (V)
V_{ACR}	Voltage drop across the air core reactor (V)
V_{c}	Control voltage (V)
V_{sfcl}	Voltage drop across the AC-SFCL (V)
V_{tc}	Voltage drop across the test coil (V)
w_{t}	Width of the *RE*BCO tape (mm)
X_{ACR}	Reactance of the air core reactor (Ω)
X_{fb}	Feedback Reactance in PHIL systems (Ω)
X_{n}	Reactance of the AC-SFCL during normal operation (Ω)
$X_{\sigma\text{p}}$	Primary stray reactance (Ω)

$X'_{\sigma s}$	Secondary stray reactance (referred to primary side) (Ω)
X_{sfcl}	Reactance of the AC-SFCL (Ω)
X_{src}	Reactance of a power source (Ω)
z_{hc}	Generated reference impedance during one half cycle (%)
Z_{hc}	Generated impedance during one half cycle (Ω)
z_{lim}	Specified reference impedance of the AC-SFCL during fault (%)
Z_{lim}	Specified impedance of the AC-SFCL during fault (Ω)
$Z_{load\text{-}A}$	Impedance of load in phase A to adjust the nominal current in PHIL simulation (Ω)
$Z_{load\text{-}B}$	Impedance of load in phase B to adjust the nominal current in PHIL simulation (Ω)
$Z_{load\text{-}C}$	Impedance of load in phase C to adjust the nominal current in PHIL simulation (Ω)
z_n	Reference impedance of the AC-SFCL during normal operation (%)
Z_n	Impedance of the AC-SFCL during normal operation (Ω)
Z_p	Impedance of the primary winding (Ω)
Z_{sfcl}	Impedance of the AC-SFCL (Ω)
Z_{source}	Impedance of the power source (Ω)
φ	Phase angle
Φ	Magnetic flux
Φ_p	Magnetic flux of the primary winding
Φ_{ps}	Magnetic flux in the primary winding generated by the secondary winding
Φ_s	Magnetic flux of the secondary winding
Φ_{sp}	Magnetic flux in the secondary winding generated by the primary winding
$\Phi_{\sigma p}$	Magnetic stray flux of the primary winding
$\Phi_{\sigma s}$	Magnetic stray flux of the secondary winding
ρ_{nc}	Normal conducting resistivity ($\Omega mm^2/m$)
ρ_p	Resistivity of the conductor in the primary winding ($\Omega mm^2/m$)

F. Literature and References

[AMS00] Ashworth S.P.; Maley M.; Suenaga M. et al. (2000): Alternating current losses in YBa2Cu3O7-x coated conductors on technical substrates. *Journal of Applied Physics*, Vol. 88 (No. 5). pp. 2718-2723.

[AMS15] AMSC (2015): *www.amsc.com*. American Superconductor. http://www.amsc.com/solutions-products/hts_wire.html, zuletzt geprüft 2015.

[APY07] Ahn M.C.; Park D.K.; Yang S.E. et al. (2007): A Study on the Design of the Atabilizer of Coated Conductor for Applying to SFCL. *IEEE Transactions on Applied Superconductivity*, Vol. 17 (No. 2). pp. 1855-1858.

[AS12] Abramovitz A. und Smedley K.M. (2012): Survey of Solid-State Fault Current Limters. *IEEE Transactions on Power Electronics*, Vol. 27 (No. 6). pp. 2770-2782.

[BAC15] Bagrets N.; Augieri A.; Celentano G. et al. (2015): Investigation of ReBCO Conductor Tape Joints for Superconducting Applications. *IEEE Transactions on Applied Superconductivity*, Vol. 25 (No. 3). pp. 6602705.

[BBD11] Bock J.; Bludau M.; Dommerque R. et al. (2011): HTS Fault Current Limiters - First Commercial Devices for Distribution Level Grids in Europe. *IEEE Transactions on Applied Superconductivity*, Vol. 21 (No. 3). pp. 1202-1205.

[BBW05] Bock J.; Breuer F.; Walter H. et al. (2005): CURL 10: Development and Field-Test of a 10 kV/10 MVA Resistive Current Limiter Based on Bulk MCP-BSCCO 2212. *IEEE Transactions on Applied Superconductivity*, Vol. 15 (No. 2). pp. 1955-1960.

[BEB05] Bock J.; Elschner S.; Breuer F. et al. (2005): Field Demonstration of World Wide Largest Superconducting Fault Current Limiter and Novel Concepts. *CIRED - 18th International Conference on Electricity Distribution*. Session No 1.

[BEB05a] Bock J.; Elschner S.; Breuer F. et al. (2005): Field Demonstration of World Wide Largest Superconducting Fault Current Limiter and Novel Concepts. *CIRED, 18th International Conference on Electicity Distribution*, Session No 1.

[Bed86] Bednorz J.G. und Müller K.A. (1986): Possible high Tc superconductivity in the Ba-La-Cu-O system. *Zeitschrift für Physik B - condensed Matter*, vol. 64. p. 189-193.

[Ber11] Berger A. (2011): *Entwicklung supraleitender, strombegrenzender Transforamtoren*. Karlsruher Schriftenreihe zur Superaleitung - Band 003. Aufl. Karlsruhe: KIT Scientific Publishing 2011, ISBN 978-3-86644-637-3.

[Ber87] Berlicourt T.G. (1987): Type II Superconductivity: Quest for Understanding. *IEEE Transactions on Magnetics*, Vol. MAG-23 (No. 2). pp. 403-412.

[BFP91] Bashkirov Y.A.; Fleischmann L.S.; Patsayeva T.Y. et al. (1991): Current-Limiting Reactro based on high-Tc Superconductors. *IEEE Transactions on Magnetics*, Vol. 27 (No. 2). pp. 1089-1092.

[BH09] Baarsma B.E. und Hop J.P. (2009): Pricing power outages in the Netherlands. *Eenrgy*, Vol. 34 (No. 9). pp. 1378-1386.

[BHK11] Bock J.; Hobl A.; Krämer S. et al. (2011): Nexans' Superconducting Fault Current Limiters for Medium Voltage Applications - Status and Prospects. *CIRED 21st International Conference on Electricity Distribution*. Paper No. 0352.

[BHS15] Bock J.; Hobl A.; Schramm J. et al. (2015): Resistive Superconducting Fault Current Limiters are Becoming a Mature Technology. *IEEE Transactions on Applied Superconductivity*, Vol. 25 (No. 3). pp. 5600604.

[BI93] Brandt E.H.a.I.M. (1993): Type-II-superconductor strip with current in a perpendicular field. *Physical Review B*, Vol. 48 (No. 17). 893-906.

[BNK11] Berger A.; Noe M. und Kudymow A. (2011): Recovery Characteristic of Coated Conductors for Superconducting Fault Current Limiters. *IEEE Transactions on Applied Superconductivity*, Vol. 21 (No. 3). pp. 1315-1318.

[BOS09] Baldan C.A.; Oliveira U.R.; Shigue C.Y. et al. (2009): Evaluation of Electrical Properties of Lap Joints for BSCCO and YBCO tapes. *IEEE Transactions on Applied Superconductivity*, Vol. 19 (No. 3). pp. 2831-2834.

[Bra94] Brand E.H. (1994): Thin superconductors in a perpendicular magnetic ac field: General formualtion and strip geometry. *Physical Review B*, Vol. 49 (No. 13). 9024-9040.

[Bru15] EST B. (2015): *www.bruker.com*. Bruker EST. https://www.bruker.com/products/superconductors-and-metal-composite-materials/superconductors.html, zuletzt geprüft 2015.

[BSS97] Baltzer E.; Schmitt H. und Schultz W. (1997): Operation of Superconducting Fault Current Limiters in Electrical Networks. *IEE Conference Publication: Eletricity Distribution. Part 1: Contributions. CIRED.*, Vol. 4 (No. 438), INSPEC Nr.: 5658744. pp. 4.5.1-4.5.6.

[Cav98] Cave J.R. (1998): Critical temperature. In Seeber B. (Hg.), *Handbook of applied superconductivity*. Bristol: Inst. of Physics Publ. p. 281-294.

[CIG12] Schmitt H.; Amon Filho J.; Adapa R. et al. (2012): *Application and Feasibility of Fault Current Limiters in Power Systems*. CIGRE Working Group A3.23, ISBN: 978-2-85873-189-3.

[CWN97] Cave J.R.; Willén D.W.A.; Nadi R. et al. (1997): Testing and Modelling of Inductive Sueprconducting Fault Current Limiters. *IEEE TRansactions on Applied Superconductivity*, Vol. 7 (No. 2). pp. 832-835.

[DCML88] Dimos D.; Chaudari P.; Mannhart J. et al. (1988): Orientation Dependance of Grain-Boundary Critical Currents in YBa2Cu3O7-d Bicrystals. *Physical Review Letter*, Vol. 61 (No. 2). pp. 219-222.

[DGL05] Duckworth R.C.; Gouge M.J.; Lue J.W. et al. (2005): Substrate and stabilization effects on the transport AC losses in YBCO coated conductors. *IEEE Transactions on Applied Superconductivity*, Vol. 15 (No.2). pp. 1583-1586.

[DKH10] Dommerque R.; Krämer S.; Hobl A. et al. (2010): First commercial medium voltage superconducting fault-current limiters: production, test and isntallation. *Supercond. Sci. Technol.*, Vol.23 (No. 3). 034020.

[DSB11] Dione M.; Sirios F. und Bonnard C.-H. (2011): Evaluation of the Impact of Superconducting Fault Current Limters on Power System Network Protections using a RTS-PHIL Methodology. *IEEE Transactions on Applied Superconductivity*, Vol. 23 (Issue 3). pp. 2193-2196.

[DSH10] Dedicatoria M.J.; Shin H.S.; Ha H.S. et al. (2010): Electro-mechanical Property Evaluation of REBCO Coated Conductor Tape with Stainless Steel Substrate. *Superconductivity and Cryogenics*, Vol. 12 (No. 4). pp. 20-23.

[EBN03] Elschner S.; Breuer F.; Noe M. et al. (2003): Manufacturing and testing of MCP 2212 bifilar coils for a 10 MVA fault current limiter. *IEEE Transactions on Apllied Superconductivity*, Vol. 13 (No. 2). pp. 1980-1983.

[EBW01] Elschner S.; Breuer F.; Wolf A. et al. (2001): Characteristics of BSCCO 2212 bulk material for resistive current limiters. *IEEE Transactions on Applied Superconductivity*, Vol. 11 (No. 1). pp. 2507-2510.

[EKB12] Elschner S.; Kudymow A.; Brand J. et al. (2012): ENSYSTROP. Design, manufacturing and test of a 3-phase resisitive fault current limiter based on coated conductors for medium voltage application. *Physica C*, Vol. 482. pp. 98-104.

[EPRI09] EPRI (2009): *Superconducting Fault Current Limters*. (Technology Watch) Electric Power Research Institute (EPRI).

[FBA93] Fleischmann L.S.; Bashkirov Y.A.; Aresteanu V.A. et al. (1993): Design Considerations for an Inductive high-Tc Superconducting Fault Current Limiter. *IEEE Transactions on Applied Superconductivity*, Vol. 3 (No. 1). pp. 570-573.

[Fle10a] Fleisch D. (2010): *A Student's Duide to Maxwell's Equations, pp. 48*. Cambridge, New York, Melburne, Madrid, Cape Town, Singapore, Sao Paulo, Delhi, Dubai, Tokyo, Mexiko City: Cambridge University Press, ISBN 978-0-521-70147-1.

[Fle10b] Fleisch D. (2010): *A student's Guide to Maxwell's Equations, pp. 116*. Camebridge, New York, Melbourne, Madrid, Cape Town, Singapore, Sao Paulo, Delhi, Dubai, Tokyo Mexiko City: Cambridge University Press, ISBN 978-0-521-7014-1.

[FNB07] Fushiki K.; Nitta T.; Baba J. et al. (2007): Design and Basic Test of SFCL of Transformer Type by Use of Ag Sheathed BSCCO Wire. *IEEE Transactions on Applied Superconductivity*, Vol. 17 (No. 2). pp. 1815-1818.

[Foe03] Foellinger O. (2003): *Laplac-, Fourier- und z-Transformation*. 8. ueberarbeitete Auflage. Aufl. Heidelberg: Huethig Verlag, ISBN 3-7785-2911-0.

[FSN99] Fujikawa K.; Shirai Y.; Nitta T. et al. (1999): Experimental Study on Adjusability of Superconducting Fault Current Limiter with Adjustible Trigger Current Level. *IEEE Transactions on Applied Superconductivity*, Vol. 9 (No. 2). pp. 1351-1354.

[Fuj15] Fujikura (2015): *www.fujikura.com*. Fujikura Ltd. http://www.fujikura.com/solutions/superconductingwire/, zuletzt geprüft 2015.

[GAK01] Groves J.R.; Arendt P.N.; Kung H. et al. (2001): Texture Development in IBAD MgO Films as a Function of Deposistion Thickness and Rate. *IEEE Transactions on Apllied Superconductivity*, Vol. 11 (No. 1). pp. 2822-2825.

[GCK14] L. G.; T. C.; J. K. et al. (2014): Finite Element Modelof a Superconducting Fault Current Limiter Calibrated by Hardware-in-the-Loop Measurements. *IEEE Transactions on Applied Superconductivity*, Vol. 24 (Issue 3). pp. 1-5.

[GF78] Gray K.E. und Fowler D.E. (1978): A superconductor fault-current limiter. *J.Appl. Phys.* (49(4)). 2546-2550.

[GR93] Giese R.F. und Runde M. (1993): Assesment Study of Superconducting Fault-Current Limiters Operating at 77 K. *IEEE TRansactions on Power Delivery*, No. 3. Vol. 8.

[Gri14] Griffiths D.J. (2014): *Introduction to Electrodynamics*. Fourth Edition. Aufl. Pearson Education Limited, ISBN: 978-1-292-02142-3.

[Haz14] Hazelton D.W. (May 2014): *SuperPower 2G HTS Conductor*. (Presentation) 1st Workshop on Accalarator Magnets in HTS - Hamburg, Germany: SuperPower.

[HDB05] Hawley C.F.; Darmann F. und Beales T.P. (2005): Performance of a 1 MVA high temperature superconductors-enabled saturable magnetic core-type fault current limiter. *Supercond. Sci. Technol.*

[HE06] Hamachi LaCommare K. und Eto J.H. (2006): Cost of power interuptions to electricity in the United States. *Energy*, Vol. 31 (No. 12). pp. 1845-1855.

[Heu07a] Heuck K.; Dettmann K.-D. und Schulz D. (2007): *Elektrische Energieversorgung: Erzeugung, Übertragung und Verteilung elektrischer Energie für Studium und Praxis*. 7. vollständig überarbeitete und erweitere Auflage. Aufl. Wiesbaden: Vieweg+Teubner.

[Hie11] Hieringer J. (2011): *Charakterisierung von YBCO Bandleitern für einen supraleitenden Strombegrenzer*. (Bachelorarbeit) Karlsruhe: Institut für Technische Physik (ITEP).

[HNC04] Hatta H.; Nitta T.; Chiba M. et al. (2004): Experimental Study on the Relationship Between the Trigger Current of Transformer-Type Superconducting Fautl Current Limiter and the Fault Phase. *IEEE Transactions on Applied Superconductivity*, Vol. 14 (No. 2). pp. 819-822.

[HXS10] Hazelton D.; Xie Y.-Y.; Selvamanickam V. et al. (2010): High-Performance 2G HTS Wire for Efficient and Reliable Electicity Supply. *IEEE Conference on Innovative Technologies for an Efficient and Reliable Electricity Supply (CITRES)*. pp. 42-46.

[HYY11] Hyun O.B.; Yim S.W.; Yu S.D. et al. (2011): Long-Term Operation and Fault Tests of a 22.9 kV Hybrid SFCL in the KEPCO Test Grid. *IEEE Transactions on Applied Superconductivity*, Vol. 21 (No. 3). pp. 2131-2134.

[IO95] Ichikawa M. und Okazaki M. (1995): A Magnetic Shielding Type Superconducting Fault Current Limiter using a Bi2212 Thick Film Cylinder. *IEEE Transactions on Applied Superconductivity*, Vol. 5 (No. 2). pp. 1067-1070.

[JGV11] Jong E.d.; Graff R.d.; Vaessen P. et al. (2011): *European Whitebook on Real-Time Powerhardware-in-the-loop*. (White Book) DERlab.

[JKK07] Janowski T.; Kozak S.; Kondratowicz-Kucewicz B. et al. (2007): Analysis of Transfoer Type Superconducting Fault Curren Limiters. *IEEE Transactions on Applied Superconductivity*, Vol. 17 (No. 2). pp. 1788-1790.

[JKM03] Janowski T.; Kozak S.; Malinowski H. et al. (2003): Properties Comparison of Superconducting Fault Current Limiters With Closed and Open Core. *IEEE Transactions on Applied Superconductivity*, Vol. 13 (No. 2). pp. 2072-2075.

[KHS63] Kim Y.B.; Hempstead C.F. und Strnad A.R. (1963): Magnetization and Critical Supercurrents. *Phys. Rev.*, Vol. 129. pp. 528.

[KI97] Kado H. und Ichikawa M. (1997): Performance of a High-Tc Superconducting Fault Current Limiter. *IEEE Transactions on Applied Superconductivity*, Vol. 7 (No. 2). pp. 993-996.

[KJK05] Kozak J.; Janowski T.; Kozak S. et al. (2005): Design and Testing of 230 V Inductive Type of Superconducting Fault Current Limiter With an Open Core. *IEEE Transactions on Applied Superconductivity*, Vol. 15 (No. 2). pp. 2031-2034.

[KJW10] Kozak S.; Janowski T.; Wojtasewicz G. et al. (2010): The 15 kV Class Inductive SFCL. *IEEE Transactions on Applied Superconductivity*, Vol. 20 (No. 3). pp. 1203-1206.

[KMJ11] Kozak J.; Majka M.; Janowski T. et al. (2011): Tests and Performance Analysis of Coreless Inductive HTS Fault Current Limiters. *IEEE Transactions on Applied Superconductivity*, Vol. 21 (No. 3). pp. 1303-1306.

[KMK12] Kozak J.; Majka M.; Kozak S. et al. (2012): Design and Tests of Coreless Inductive Superconducting Fault Current Limiter. *IEEE Transactions on Applied Superconductivity*, Vol. 22 (No. 3). pp. 5601804.

[KMK13] Kozak J.; Majka M.; Kozak S. et al. (2013): Comparison of Inductive and Resistive SFCL. *IEEE Transactions on Applied Superconductivity*, Vol. 23 (No. 3). pp. 5600604.

[Kom95] Komarek P. (1995): *Hochstromanwendung der Supraleitung*. Stuttgart: Teubner, ISBN 3-519-03225-2.

[KSD12] Kapoor R.; Shukla A. und Demetriades G. (2012): State of Art of Power Electronics in Circuit Breaker Technology. *IEEE Energy Conversion Congress and Exposition (ECCE)*. pp. 615-622.

[KYT05] Kovalsky L.; Yuan X.; Tekletsadik K. et al. (2005): Appliacations of Superconducting Fault Current Limiters in Electric Power Transmission Systems. *IEEE Transactions on Applied Superconductivity*, Vol. 15 (No. 2). pp. 2130-2133.

[LBS03] Lakner M.; Braun D.; Schnyder G. et al. (2003): *Marktpotential von supraleitenden Strombegrenzern*. (Schlussbericht) Bundesamt für Energie, DIS-Projekt Nr. 44237; DIS-Vertrags Nr. 84281.

[Lei06] Leibfried P.D.-I.T. (WS2006 / SS 2007): *Elektrische Anlagen- und Systemtechnik*. Manuskript zur Vorlesung. Aufl. Karlsruhe: Institut für Elektroenergiesysteme und Hochspannungstechnik.

[LGFP01] Larbalestier D.; Gurevich A.; Feldmann D.M. et al. (2001): High-Tc supercnducting materials for electric power applications. *Nature*, Vol. 414. pp. 368-377.

[LGX07] Li L.F.; Gong L.H.; Xu X.D. et al. (2007): Field Test and Demosntrated Operation of 10.5 kV/1.5 kA Fault Current Limiter. *IEEE Transactions on Applied Superconductivity*, Vol. 17 (No. 2). pp. 2055-2058.

[LHS11] Lu J.; Han K.; Sheppard W.R. et al. (2011): Lap Joint Resistance of YBCO Coated Conductors. *IEEE Transactions on Applied Superconductivity*, Vol. 21 (No. 3). pp. 3009-3012.

[LT11] Leahy E. und Tol R.S.J. (2011): An estimate of the value of lost load for Ireland. *Energy Policy*, Vol. 39 (No. 5). pp. 1514-1520.

[Mae88] Maeda H.; Tanaka Y.; Fuutomi M. et al. (1988): New high-Tc superconductors without rare earth element. *Physica C: Superconductivity 153-155* (Part 1). pp. 602-607.

[Mar99] Marlene M. (1999): *Wechselstromtechnik: Grundlagen und Beispiele*. Vieweg Verlag, ISBN 3-528-07437.

[MBA12] Martini L.; Bocchi M.; Ascade M. et al. (2012): Development, testing and isntallation of a Superconducting Fault Current Limiter for medium voltage distribution networks. *Physics Procedia*, Vol. 36. pp. 914-920.

[MBA13] Martini L.; Bocchi M.; Ascade M. et al. (2013): The first Italian Superconducting Fault Current Limiter: Results of the field testting experience after one year operation. *Journal of Physics: Conference Series - 11th European Conference on Applied Superconductivity (EUCAS)*, Vol. 507. pp. 032003.

[MBA15] Martini L.; Bocchi M.; Angeli G. et al. (2015): Live Grid Field-Testing Final Result of the First Italian Superconducting Fault Current Limiter and Severe 3-Phase Fault Experience. *IEEE Transactions on Applied Superconductivity*, No. 3. Vol. 25.

[MD09] Moscrop J. und Darmann F. (2009): Design and development of a 3-phase saturated core High Temperature Superconducting Fault Curent Limiter. *International Conference on Electric Power and Energy Conversion Systems (EPECS)*. pp. 1-6.

[Mil11] Millman K. Jarrod A.M. (2011): Python for Scientists and Engineers. *Computing in Science & Engineering* (13). 9-12.

[MKK15] Majka M.; Kozak J.; Kozak S. et al. (2015): Design and Numerical Analysis of the 15 kV Class Coreless Inductive Type SFCL. *IEEE Transactions on Applied Superconductivity*, Vol. 25 (No. 3). pp. 5601005.

[MO12] Meissner W. und Ochsenfeld R. (1912): Ein neuer Effekt bei Eintritt der Supraleiterfähigkeit. *Die Naturwissenschaften*, 21. 44.

[Moo14] Moon S.H. (May 2014): *HTS Development and Industrialization at SuNAM*. (Presentation) 1st Workshop on Accelerator Magnets in HTS - Hamburg, Germany: SuNAM.

[Mor13] Morandi A. (2013): State of the art of superconducting fault current limiters and their application to the electric power system. *Physica C*, Vol. 484. pp. 242-247.

[MRD11] Moriconi F.; De La Rosa F.; Darnmann F. et al. (2011): Development and Deployment of Saturated-Core Fault Current Limiters in Distribution and Transmission Substations. *IEEE Transactions on Applied Superconductivity*, Vol. 21 (No. 3). pp. 1288-1293.

[MSD04] Meyer C.; Schroeder S. und De Doncker R.W. (2004): Solid-State Cirucuit Breakers and Current Limiters for Medium-Voltage Systems Having Distrbuted Power Systems. *IEEE Transactions on Power Electronics*, Vol. 19 (No. 5). pp. 1333-1340.

[MSG95] Meerovich V.; Sokolovsky V.; Grzegorz J. et al. (1995): High-Tc Superconducting Inductive Current Limiter for 1kV/25A Performance. *IEEE Transactions on Applied Suerpconductivity*, No. 2. Vol. 5.

[MW01] Magnusson N. und A. W. (2001): AC losses in high tmeperature superconducting tapes exposed to longitudinal magnetic fields. *Cryogenics*, Vol. 41. 721-724.

[MZC15] Miao Q.; Zhu J.M.; Cheng M. et al. (2015): Fabrication and Characteristic Tests of a Novel Low-Resistance Joint Structure for YBCO Coated Conductors. *IEEE Transactions on Applied Superconductivity*, Vol. 25 (No. 3). pp. 6600705.

[Nae10] Naeckel O. (2010): *Supraleitender Magnetischer Energiespeicher zur Kompensation fluktuierender erneuerbarer Energieerzeugung*. (Diplomarbeit) Karlsruhe: Institut for Technical Physics.

[NAJ04] Nishioka T.; Amemiya N.; Jiang Z. et al. (2004): Influence of silver layer thickness on magnetization loss of YBCO coated conductors in transverse magnetic field with various orientations. *Physica C*, Vol. 412-414. pp. 992-998.

[NB88] Namjoshi K.V. und Biringer P.P. (1988): Low-Frequency Eddy-Current Loss estimation in Long Conductors by Using the Moment of Inertia of Cross Sections. *IEEE Transactions on Magnetics*, 24 (5).

[NKB07] Nooji M.d.; Koopmans C. und Bijvoet C. (2007): The value of supply security The costsof power interuptions: Economic input for damage reduрction and investment networks. *Energy Econimics*, Vol. 29 (No. 2). pp. 277-295.

[NO99] Noe M. und Oswald B.R. (1999): Technical and Economical Benefits of Superconducting Fault Current Limiters in Power Systems. *IEEE Transactions on Applied Superconductivity*, Vol. 9 (No. 2). pp. 1347-1350.

[NP97] Norris W.T. und Powers A. (1997): Fault current limiters using superconductors. *Cryogenics*, Vol. 37 (No. 10). pp. 657-665.

[NS07] Noe M. und Steurer M. (2007): High-temperature superconductor fault current limiters: concepts, applications and development status. *Supercond. Sci. Technol.* (20). 15-27.

[NSK09] Neumueller H.W.; Schimidt W.; Kraemer H.P. et al. (2009): Development of Resistive Fautl Current Limiters Based on YBCO Coated Conductors. *IEEE Transactions on Applied Superconductivity*, Vol. 19 (No. 3). pp. 1950-1955.

[NSS09] Nii T.; Shouno Y. und Shirai Y. (2009): Basic Experiments on Transformer Type SCFCL of Rewound Structure Using BSCCO Wire. *IEEE Transactions on Applied Superconductivity*, Vol. 19 (No. 3). pp. 1892-1895.

[ONN11] Oda S.; Noda S.; Nishioka H. et al. (2011): Current Limiting Experiment of Transformer Type Superconducting Fault Current Limiter With Rewound Stgructure Using BSCCO Wire in Small Model Power System. *IEEE Transactions on Applied Superconductivity*, Vol. 21 (No. 3). pp. 1307-1310.

[Onn12] Onnes H.K. (1912): Further experiments with Liquid Helium G. On the electrical resistance of Pure Metals etc. VI. On the Sudden Change in the Rate at which the Resistance of Mercury Disappears. *KNAW Proceedings 14 II.*, Amsterdam: Huygens Institute. p. 818-821.

[ONO04] Ogawa J.; Nakayama H.; Odaka S. et al. (2004): AC loss characteristics of YBCO coated conductors with Ag protection layer. *Physica C*, Vol. 412-414 Part 2. pp. 1021-1025.

[PAK07] Park D.K.; Ahn M.C.; Kim H.M. et al. (2007): Analysis of a Joint Method Between Superconducting Coated Conductors. *IEEE Transactions on Applied Superconductivity*, Vol. 17 (No. 2). pp. 3266-3269.

[PE13] Paran S. und Edrington C.S. (2013): Improved Power Hardware-in-the-Loop interface methods via impedance matching. *IEEE / Electric Ship Technologoes Symposium (ESTS)*. pp. 342-346.

[PHE11] Praktiknjo A.J.; Hähnel A. und Erdmann G. (2011): Assessing energy supply security: Outage costs in private households. *Energy Policy*, Vol. 39 (No. 12). pp. 7825-7833.

[Phi00] Philippow E. (2000): *Grundlagen der Elektrotechnik*. 10. durchgesehene und ergänzte Auflage. Aufl. Berlin: Verlag Technik Berlin, ISBN 3-341-01241-0.

[Pis09] Pistora G. (2009): *Berechnung von Kurzschlussströmen und Spannungsfällen*. VDE-Schriftenreihe Normen verständlilch. Aufl. Berlin und Offenbach: VDE Verlag GmbH.

[PLR97] Paul W.; Laker M.; Rhyner J. et al. (1997): Test of 1.2 MVA high-Tc superconducting fault current limiter. *Supercond. Sci. Technol.*, Vol. 19 (No. 12). pp. 914-918.

[Pre83] Prechtl A. (1983): Felder und Kräfte in Zylinderspulen. *Archiv für Elektrotechnik*, Vol. 66 (No. 5-6). pp. 351-364.

[RFW07] Rozenshtein V.; Friedman A.; Wolfus Y. et al. (2007): Saturated Cores FCL - A New Approach. *IEEE Transactions on Applied Superconductivity*, Vol. 17 (No. 2). pp. 1756-1759.

[RIA13] Ratha A.; E. I. und Andersson G. (2013): Value of Lost Load: How much is supply security worth. *IEEE Power and Energy Society General Meeting (PES)*, INSPEC: 13933036.

[RLS13] Rupich M.W.; Li X.; Sathyamurthy S. et al. (2013): Second Generation Wire Development at AMSC. *IEEE Transaction on Applied Superconductivity*, Vol. 23 (No. 3). pp. 6601205.

[RLT09] Rupich M.W..X.L.; Thieme C.; Sathyamurthy S. et al. (2009): Advances in second generation high temperature superconducting wire manufacturing and R&D at American Superconductor Corporation. *Superconducting Science and Technology* (No. 23).

[RPB82] Raju B.P.; Parton K.C. und Bartram T.C. (1982): A Current Limiting Device Using Superconducting D.C. Bias - Applications and Prospects. *IEEE Transactions on Power Apparatus and Systems*, Vol. PAS-101 (No. 9). pp. 3173-3177.

[RSB08] Ren W.; Steurer M. und Baldwin T.L. (2008): Improve the stability and the accuracy pf Power Hardware-in-the-Loop Simualtions by Selecting Appropriate Interface Algorithms. *IEEE Transactions on Industry Applications*, Vol. 44 (Issue 4). pp. 1286-1294.

[RSS12] Reichl J.; Schmidthaler M. und F. S. (2012): The value of supply security: The cost of power outages to Austrian houselolds, firms and the public sector. *Energy Economics*, Vol. 36. pp. 256-261.

[SBN08] Suzuki K.; Baba J. und Nitta T. (2008): Conceptual design of an SFCL by use of BSCCO wire. *Journal of Physics: Conf. Ser.*, Vol. 97 (No. 1). pp. 012293.

[Scha09] Schacherer C. (2009): *Theoretische und experimentelle Untersuchungen zur Entwicklung supraleitender resistiver Strombegrenzer*. Karlsruher Schriftenreihe zur Supratleitung Band 001. Aufl. Karlsruhe: Universitätsverlag Karlsruhe, ISBN: 978-3-86644-412-6.

[Scha13] Schacherer C. (2013). (personal communication) Karlsruhe: KIT.

[Schl05] Schlabbach J. (2005): *Short-circuit Currents*. London: The Institution of Electrical Engineers.

[Schw02] Schwab A.J. (2002): *Begriffswelt der Feldtheorie: Elektrimagnetische Felder, Maxwellsche Gleichungen, Gradient, Rotation, Divergenz. pp. 109*. Berlin, Heidelberg, New York: Springer Verlag, ISBN 3-540-42018-5.

[Schw06] Schwab A.J. (2006): *Elektroenergiesysteme*. Berlin, Heidelberg: Springer-Verlag.

[SCK11] Selvamanickam V.; Chen Y.; Kesgin I. et al. (2011): Progress in Performance Improvement and New Research Areas for Cost Reduction of 2G HTS Wires. *IEEE Transactions on Applied Superconductivity*, Vol. 21 (No. 3). pp. 3049-3054.

[SCX08] Selvamanickam V.; Chen Y.; Xiong X. et al. (2008): Progress in second-generation HTS wire development and manufacturing. *Physica C: Superconductivity and its Applications*, Vol. 468 (Issues. 15-20). pp. 1504-1509.

[SCX09] Selvamanickam V.; Chen Y.; Xiong X. et al. (2009): High Performance 2G Wires: From R&D to Pilot-Scale Manufacturing. *IEEE Transactions on Applied Superconductivity*, Vol. 19 (No. 3). pp. 3225-3230.

[SD10] Shin H.-S. und Dedicatoria M.J. (2010): Comparison of the Bending Strain Effect on Transport Property in Lap- and Butt-Jointed Coated Conductor Tapes. *IEEE Transactions on Applied Superconductivity*, Vol. 20 (No. 3). pp. 1541-1544.

[SDG14] Shin H.S.; Dedictoria M.J.; Goropse A. et al. (2014): Strain and magnetic field response of Ic on reinforced GdBCO coatred conductor tapes at 77 K. *AIP Conference Proceedings 1574, 239*.

[Sel10] Selvamanickam V. und al. e. (August 2010): *Progress in research and development of IBAD-MOCVD based spuerconducting wires.* (Presentation) Applied Superconductivity Conference, Washington D.C.: University of Houston.

[Sel11] Selvamanickam V. (February 2011): *Second-generation HTS Wire for Wind Energy Applications.* (Presentation) Symposium on Superconducting Devices for Wind Eenrgy - Barcelona, Spain: University of Houston.

[SES10] Steuerer M.; Edrington C.S.; Sloderbeck W. et al. (2010): A Megawatt-Scale Power Hardware-in-the-Loop Simulation Setup for Motor Drives. *IEEE Transactions on Industiral Electronics*, Vol. 57 (Issue 4). pp. 1254-1260.

[SFH99] Shirai Y.; Fujikawa K.; Hagiwara K. et al. (1999): Recovery Characteristics of Fault Current Limiter with Adjustible Trigger Current Level. *IEEE Transactions on Applied Superconductivity*, Vol. 9 (No. 2). pp. 1381-1384.

[SID95] Salasoo L.; Imece A.F.; Delmerico R.W. et al. (1995): Comparison of superconducting fault limiter concets in electric utiliy applications. *IEEE Transactions on Applied Superconductivity*, No. 2. Vol. 5.

[SKN07] Schmidt W.; Kraemer H.P.; Neumuller H.W. et al. (2007): Investigation of YBCO Coated Conductors for FAult Current Limiter Applications. *IEEE Transactions on Applied Superconductivity*, Vol. 17 (No. 2). pp. 3471-3474.

[SLS09] Schacherer C. L.J.,S.M.,N.M: (2009): Power Hardware in the Loop Testing of a YBCO Coated Conductor Fautl Current Limiter Module. *IEEE Transactions of Applied Superconductivity*, Vol. 19 (Issue 3 Part 2). pp. 1801-1805.

[SMC10] Sokolowsky V.; Meerovich V.; Chubraeva L.I. et al. (2010): An improved design of an inductive fault current limiter based on a superconducting cylinder. *Supercond. Sci. Technol.*, Vol. 23 (No. 8). pp. 085007.

[SMV04] Sokolovsky V.; Meerovich V.; Vajda I. et al. (2004): Superconducting FCL: Design and Application. *IEEE Transactions on Applied Superconductivity*, Vol. 14 (No. 3). pp. 1990-2000.

[SNO09] Shirai Y.; Nii T.; Oda S. et al. (2009): Current Limiting Characteristics of Transformer Type HTS Superconducting Fault Current Limiter with Rewound Structure. *Journal of Physics: Conference Series*, Vol. 234 (No. 3). pp. 032053.

[SNY12] Shirai Y.; Noda S.; Yamabe K. et al. (2012): Current limiting perfrmance test of 3-phase tri-axial transformer-type SFCL with re-wound structure at 3-line-to-ground fault in lab-scale transmission system. *Physica C*, Vol. 484. pp. 248-252.

[SNY13] Shirai Y.; Noda S.; Yamabe K. et al. (2013): Current Limiting Performance of Three-Phase Transformer Type SFCL at Unbalanced Fault Conditions. *IEEE Transactions on Applied Superconductivity*, Vol. 23 (No. 3). pp. 5601905.

[SOX15] SOX (2015): *www.superox.ru.* SuperOX. http://www.superox.ru/en/products/42-2G-HTS-tape/, zuletzt geprüft 2015.

[SP15] SP (2015): *www.superpower-inc.com.* SuperPower. http://www.superpower-inc.com/system/files/SP_2G+Wire+Spec+Sheet_2014_web_v1_0.pdf, zuletzt geprüft 2015.

[Str06] Strassacker G. und Roland S. (2006): *Rotation, Divergenz und Gradient: Einführung in die elektromagnetische Feldtheorie, pp.44.* 6., durchgesehene und ergänzte Auflage. Aufl. Wiesbaden: Teubner Verlag, ISBN 3-8351-0048-3.

[SuN15] SuNam (2015): *www.i-sunam.com.* SuNam. http://www.i-sunam.com/down/SuNAM_2G_Wire_Spec.pdf, zuletzt geprüft 2015.

[Sup12] Superpower (2012): *www.superpower-inc.com.* www.superpower-inc.com,

[Tra07] Oliphant T.E. (2007): Python for Scientific Computing. *Computing in Science & Engineering* (9). 10-20.

[UMD09] Usoskin A.; Mumford F.; Dietrich R. et al. (2009): Inductive Fault Current Limiters: Kinetics of Quenching and Recovery. *IEEE Transactions on Applied Superconductivity*, No. 3. Vol. 19.

[WAT87] Wu M.K.; Ashburn J.R.; Torng C.J. et al. (1987): Superconductivity at 93 K in a new mixed-phase Y-Ba-Cu-O compound system at ambient pressure. *Phys. Rev. Lett.* (No. 9). pp. 908.

[WJK14] Wojtasevicz G.; Janokowski T.; Kozak S. et al. (2014): Experimantal Investigation of a Model of a Transformer-Type Superconducting Fault Current Limiter With a Superconducting Coil Made of a 2G HTS Tape. *IEEE Transactions on Applied Superconductivity*, Vol. 24 (No. 3). Art. No. 5601005.

[XGS13] Xin Y.; Gong W.Z.; Sun Y.W. et al. (2013): Factory and Field Tests of a 220 kV/300 MVA Saturated Iron-Core Superconducting Fault Current Limiter. *IEEE Transactions on Applied Superconductivity*, Vol. 23 (No. 3). pp. 5602305.

[XHW11] Xin Y.; Hong H.; Wang J.Z. et al. (2011): Performance of the 35 kV/90 MVA SFCL Live-Grid Fault Current Limiting Tests. *IEEE Transcations on Applied Superconductivity*, No. 3. Vol. 21.

[XKZ09] Xiong X.; Kim S.; Zdun K. et al. (2009): Progress in High Throughput Processing of Long-Length, High Quality, abd Low Cost IBAD MgO Buffer TTapes at Superpower. *IEEE Transactions on Applied Superconductivity*, Vol. 19 (No. 3). pp. 3319-3322.

[XSM09] Xie Y.; S.V.; Marchevsky M. et al. (2009): Second-Generation HTS Wire Manufacturing and Technology Advancement at SuperPower. *IEEE Proceedings of International Converence on Apllied Superconductivity*. pp. 398-402.

[YKM09] Yazawa T.; Koyanagi K.; Takahashi M. et al. (2009): Design and Experimental Results of Three-Phase Superconducting Fault Current Limiter Using Highly-Resistive YBCO Tapes. *IEEE Transactions on Applied Superconductivity*, Vol. 19 (No. 3). pp. 1956-1959.

[YYS15] Yonemura N.; Yamabe K.; Shirai Y. et al. (2015): Current Limitng Performance of Transformer-Type Supercondcuting Fault Current Limiter Made of BSCCO and REBCO Wires. *IEEE Transactions on Applied Superconductivity*, Vol. 25 (No. 3). pp. 5601104.

[ZLF11] Zhang Y.; Lehner T.F.; Fukushima T. et al. (2011): Progress in Production and Performance of Second Generation (2G) HTS Wire for Practical Applications. *IEEE Transactions on Applied Superconductivity*, Vol. 24 (No. 5). pp. 7500405.

[ZLF14] Zhang Y.; F. L.T.; Fukushima T. et al. (2014): Progress in Production and Performance of Second Generation (2G) HTS Wire for Practical Appliactions. *IEEE Transactions on Applied Superconductivity*, Vol. 24 (No. 5). pp. 7500405.

[ZP12] Zachariadis T. und Poullikkas A. (2012): The costs of power outages: A case study from Cyprus. *Energy Policy*, Vol. 51. pp. 630-641.

Karlsruher Schriftenreihe zur Supraleitung
(ISSN 1869-1765)

Herausgeber: Prof. Dr.-Ing. M. Noe, Prof. Dr. rer. nat. M. Siegel

Die Bände sind unter www.ksp.kit.edu als PDF frei verfügbar oder als Druckausgabe bestellbar.

Band 001
Christian Schacherer
Theoretische und experimentelle Untersuchungen zur Entwicklung supraleitender resistiver Strombegrenzer. 2009
ISBN 978-3-86644-412-6

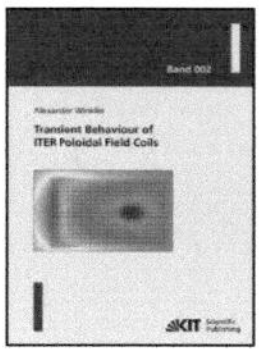

Band 002
Alexander Winkler
Transient behaviour of ITER poloidal field coils. 2011
ISBN 978-3-86644-595-6

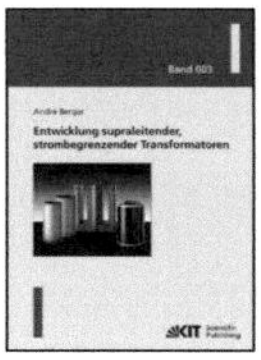

Band 003
André Berger
Entwicklung supraleitender, strombegrenzender Transformatoren. 2011
ISBN 978-3-86644-637-3

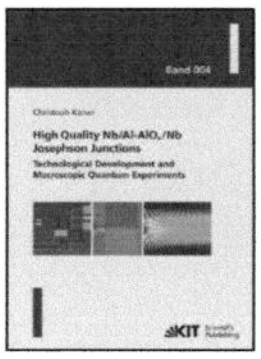

Band 004
Christoph Kaiser
High quality Nb/Al-AlOx/Nb Josephson junctions. Technological development and macroscopic quantum experiments. 2011
ISBN 978-3-86644-651-9

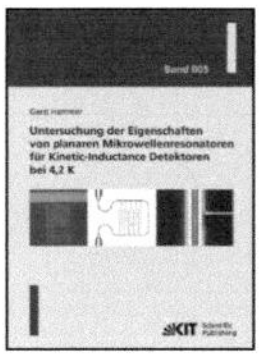

Band 005
Gerd Hammer
Untersuchung der Eigenschaften von planaren Mikrowellenresonatoren für Kinetic-Inductance Detektoren bei 4,2 K. 2011
ISBN 978-3-86644-715-8

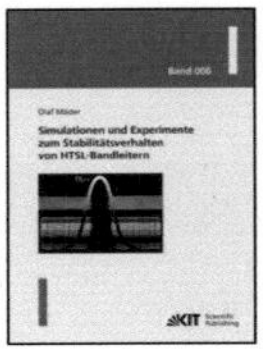

Band 006
Olaf Mäder
Simulationen und Experimente zum Stabilitätsverhalten von HTSL-Bandleitern. 2012
ISBN 978-3-86644-868-1

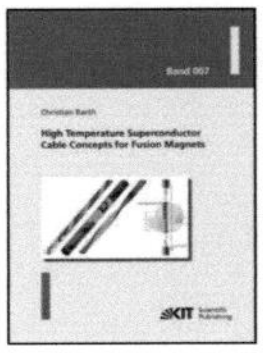

Band 007
Christian Barth
High Temperature Superconductor Cable Concepts for Fusion Magnets. 2013
ISBN 978-3-7315-0065-0

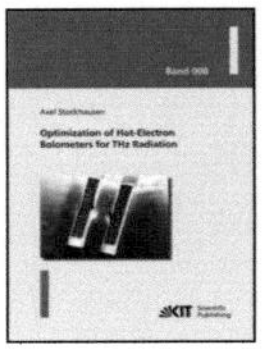

Band 008
Axel Stockhausen
Optimization of Hot-Electron Bolometers for THz Radiation. 2013
ISBN 978-3-7315-0066-7

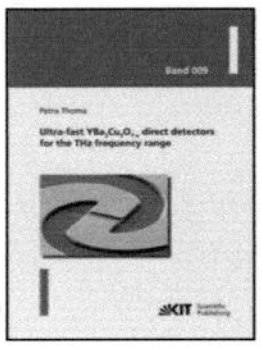

Band 009
Petra Thoma
Ultra-fast $YBa_2Cu_3O_{7-x}$ direct detectors for the THz frequency range. 2013
ISBN 978-3-7315-0070-4

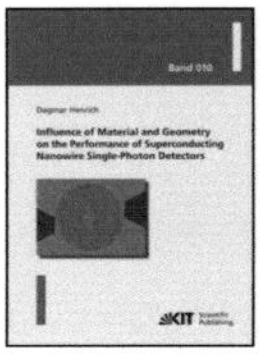

Band 010
Dagmar Henrich
Influence of Material and Geometry on the Performance of Superconducting Nanowire Single-Photon Detectors. 2013
ISBN 978-3-7315-0092-6

Band 011
Alexander Scheuring
Ultrabreitbandige Strahlungseinkopplung in THz-Detektoren. 2013
ISBN 978-3-7315-0102-2

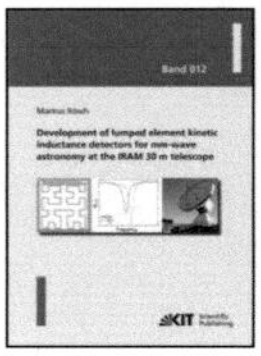

Band 012
Markus Rösch
Development of lumped element kinetic inductance detectors for mm-wave astronomy at the IRAM 30 m telescope. 2013
ISBN 978-3-7315-0110-7

Band 013
Johannes Maximilian Meckbach
Superconducting Multilayer Technology for Josephson Devices. 2013
ISBN 978-3-7315-0122-0

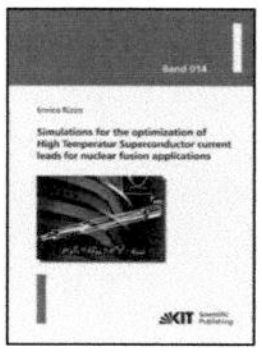

Band 014
Enrico Rizzo
Simulations for the optimization of High Temperature Superconductor current leads for nuclear fusion applications. 2014
ISBN 978-3-7315-0132-9

Band 015
Philipp Krüger
Optimisation of hysteretic losses in high-temperature superconducting wires. 2014
ISBN 978-3-7315-0185-5

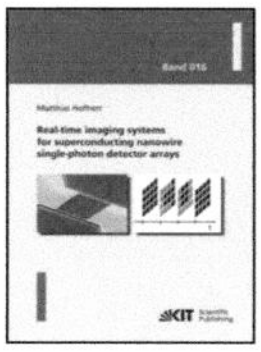

Band 016
Matthias Hofherr
Real-time imaging systems for superconducting nanowire single-photon detector arrays. 2014
ISBN 978-3-7315-0229-6

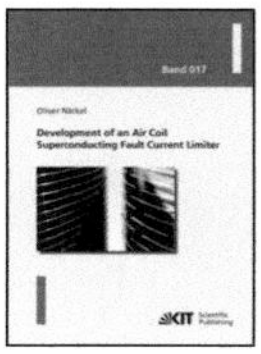

Band 017
Oliver Näckel
Development of an Air Coil Superconducting Fault Current Limiter. 2016
ISBN 978-3-7315-0526-6